CONTACT WITH SPACE

CONTACT WITH SPACE

ORANUR
SECOND REPORT 1951 – 1956

OROP DESERT Ea
1954 – 1955

by

WILHELM REICH

Haverhill House Publishing LLC
Haverhill, MA

Copyright renewed 1985 by Mary Boyd Higgins
Director of the Wilhelm Reich Infant Trust

Original © 1957
by Wilhelm Reich
Orgonon, Rangeley, Maine

This renewed edition was duplicated from the original *Core Pilot Press* edition with permission from

The Wilhelm Reich Infant Trust
P.O. Box 687
Rangeley, Maine 04970
http://www.wilhelmreichtrust.org/

All rights reserved. No part of this book may be reproduced or transmitted in any form or by any electrical or mechanical means, including photocopying, recording or by any information storage and retrieval system, without the written permission of the author or publisher, except where permitted by law.

This edition is
©2018 The Wilhelm Reich Infant Trust
Foreword ©2018 Michael Mannion

ISBN-13: 978-1-949140-95-8 (Hardcover)
ISBN-13: 978-1-949140-96-5 (Paperback)

(NOTE: Whoever is interested in studying original documents regarding Oranur, OROP Desert Ea, and the Conspiracy concerning OROP Desert Ea, is referred to *Record Appendix, Vol. III,* Library of Congress Catalog Card No. 57-8813)

Haverhill House Publishing LLC
643 E Broadway
Haverhill, MA 01830-2420

Visit us on the web at www.HaverhillHouse.com

CONTACT WITH SPACE—FOREWORD
by
Michael Mannion

Contact with Space is a challenging, compelling, and richly rewarding book. Although written under difficult, perilous circumstances, it conveys, in Reich's words, "the excitement of the adventures which open the Cosmic Age."

Published in a limited edition after Wilhelm Reich's death in a Federal prison in 1957, it was written under severe pressure from both the critical environmental conditions Reich was battling and the unrelenting attacks from conspiring political enemies and commercial interests whose goal was to destroy the newborn science of orgonomy.

The complete title of this book is *Contact with Space—ORANUR Second Report 1951-1956*. It is the continuation of the research Reich first described in *The Oranur Experiment—First Report (1947-1951)*. Those who familiarize themselves with this earlier work will have a deeper understanding of *Contact with Space*.

Reich's book is, sadly, more timely and more profoundly needed than ever. As this is being written, headlines such as the following are common: "The Northern Hemisphere Is on Fire." At present, 38% of the US is experiencing moderate to severe drought and 25% extreme drought. The four hottest years in recorded weather history have been the previous four years. And 17 of the warmest years have occurred in the last 18 years.

Coniferous forests are burning at a rate not seen in 10,000 years. Hundreds of forest fires are raging in Western Canada and the Western United States from Alaska to Southern California. The smoke from these fires has reached Europe. In Europe, forests fires are burning from the Arctic Circle to Greece. The verdant lands of the UK and Ireland appear brown in satellite photographs. It is so hot in Western Europe that many nuclear power plants are being shut down because the river water used to cool the reactors is too hot to do so.

At this moment, about 50% of the planet is experiencing drought and forests are dying on every forested continent. In addition, 21 of the Earth's 37 most important aquifers are in danger of depletion within a few years. The majority of humanity and its political leadership are not confronting this global crisis.

In this work, Reich described himself as a "Planetary Citizen" facing what he called our "Planetary Emergency." He was not understood at that time. But the crisis Reich addressed, which he felt threatened the existence of Life on Earth, has deepened dramatically in the intervening 60 years. Major environmental groups now use the term "Planetary Emergency," and people all over the world are now calling themselves "Planetary Citizens." They understand that this change in human consciousness is needed to address the global climate crisis we face.

The cover copy on the dust jacket of the original edition provided important insights into Reich's views of his last book. Readers were informed that the contents were "the result of six years of intensive research and field work" and that it was "published in relation to the current Western Dust Bowl problem... (and the) larger problem of the planetary DOR emergency." The United States faced the second worst drought in its history from 1953-1957. And moderate to extreme drought has plagued the USA and the world in every decade since.

Contact with Space provides a scientific account of the efforts to combat the drought the USA was facing then and offered information about *practical measures* that could be used to do so. Reich provides an exposition of the newest techniques of Cosmic Orgone Engineering. New, basic facts about Nature, and about a new energy technology, are introduced and explored. In addition, Reich examines the *human reactions* to these revolutionary developments, providing some background on the social milieu in which this work was undertaken.

A NEW VIEW OF THE UNIVERSE

In *The Oranur Report*, Reich wrote that "We are on the verge of an entirely new view of the universe." And Reich's view of the universe was fundamentally different from that of mechanistic science. In *Contact With Space,* that new view gained in clarity.

The science of Reich's time asserted that outer space was a vacuum. Reich's experiments led him to conclude that there is no vacuum. Rather, a physical cosmic energy permeates the universe. Today, it is the view of science that outer space is an infinite energy continuum, not a vacuum. And much scientific and technological research is underway to try to harness the power of this virtually limitless energy for space travel and for the generation of electricity.

In Reich's era, science claimed that the existence of life was an accident; too great an accident to have occurred more than once. Astrophysicists and astronomers always noted that no planets had been discovered outside of our solar system on which life could exist. Even if some planets would be found in the future, they said, the distances were so vast no one could create a technology to travel in interstellar space.

In contrast, Reich had long believed that Life, including intelligent life, existed in creation. In 1942, in *The Function of the Orgasm*, he criticized the scientific narrow-mindedness of his day, writing "You still believe that the earth, one of millions of planets, is the sole planet which bears living matter." He thought it was a scientific necessity to assume the existence of life in the cosmos. With the discovery of the orgone energy, he believed that an energy source for interstellar space travel had been discovered, one which allowed for travel beyond the speed of light.

Today, the dominant view of science is that life is present throughout the cosmos and that it began quite early in the evolution of the universe. A basic goal of NASA has always been to discover life beyond Earth, including intelligent life. In the past, they would not come right out and say it, but today they do. And the USAF and other organizations are hard at work on "Breakthrough Propulsion Projects" that they hope will make interstellar space exploration a reality.

The new orgone energy technology and new bioenergetic insights into human character structure described in this book are not known to many today. This is a gap that needs to be corrected. The knowledge Reich gained and left as his legacy needs to be investigated scientifically and developed. This is essential for us if we are to escape self-destruction and create a peaceful, loving world for generations to come.

LIFE BEYOND EARTH

The most controversial aspect of this book has been, and probably still is for many, Reich's acknowledgment that intelligent life from beyond earth has entered our lives. He took the reports from the 1950s about UFOs, or spaceships, as he preferred to call them, seriously. Many who support Reich's work were then, and still are, embarrassed by his writings on this subject. They feel it discredits, or can be used maliciously to discredit, his valuable earlier discoveries. It is my opinion that Reich was correct to take it seriously and that it is a lack of knowledge that leads advocates of his science to avoid this subject.

Why did Wilhelm Reich take the worldwide reports of UFOs in the early 1950s seriously? The main reason he did so was scientific. He saw the connection between details of the reports of anomalous aerial phenomena and the mathematical orgonometrical equations he had written between 1940 and 1944. People in all walks of life, all over the world, were reporting observations of craft operating contrary to the laws of mechanistic physics—but in accordance with the laws of orgone physics. Reich put forth the proposition that the spaceships were using orgone energy as their source of power in the atmosphere and beyond.

In Reich's time, only one other major scientist wrote and spoke openly about the UFO enigma—James E McDonald. (He was one of the first scientists to discover that chemical emissions were damaging the planet's ozone layer.). In 1953, he was invited to help establish a meteorology and atmospherics program at the University of Arizona as a professor of meteorology. McDonald eventually became the head of the Institute of Atmospheric Physics but resigned as its administrator after about a year. Reich arrived in Tucson AZ in 1954 just after McDonald left. Both openly reported having had their own UFO sightings. Both suffered as a result because scientists of that era, and government and military officials, derided, debunked and ridiculed the UFO subject. Today, more scientists are writing openly about the UFO phenomenon and extraterrestrial life. Even some governments are becoming more open about what they know, such as the UK, France, Mexico, and Chile.

We now know that from 1947 through 1952, investigators for the US Air Force had come to the conclusion that UFOs were interplanetary. A full report on this was handed to General Hoyt Vandenberg who told its authors "You have reached the wrong conclusion." He ordered all copies of this report destroyed. The CIA entered the UFO field at that time, and the official policy became one of debunking, denying, and deriding. However, Reich was influenced in his views by the publications of the few honest military voices—Major Donald E Keyhoe and Col. Edward J Ruppelt—and with good reason.

Along with the entire nation, Reich also was influenced in his views by two UFO incidents in Washington, D.C. on the weekends of July 19 and July 26, 1952, which resulted in headlines in newspapers across the country like this one in the *Washington Post*: "'Saucer Outruns Jet, Pilot Reveals." A jet pilot was sent up to investigate the objects, but he was unable to overtake the moving glowing lights. According to the *Washington Post*, the UFOs hovered only 1,700 feet above the White House lawn. (Shades of *The Day the Earth Stood Still*, a great 1951 film on UFOs that Reich writes about in this book!) An Air Force investigation was kept secret.

If we look at Reich's evaluation of the UFO phenomenon in the context of his time, we see that it makes perfect sense given the information available. And if we view his thoughts from our vantage point 60 years later, we can see that he was correct to take this phenomenon seriously. Why? To give but one example: on Sunday, December 17, 2017, *The New York Times* published a page-one article in which the US Government acknowledged publicly for the first time that it has been conducting serious investigations into the UFO phenomenon for decades. The article detailed one "black budget" project in particular. The online version included digital images of a UFO being chased by American fighter jets.

It seems worthwhile to take this look at Reich's writings on UFOs, not only because they are still controversial, but also because he once again appears to have been ahead of his time. Many scientists, especially astrophysicists, are now writing books about the UFOs and the existence of intelligent life in the universe. For over 30 years, academics, psychiatrists, historians, theologians, and other serious investigators have examined aspects of this phenomenon that Reich first addressed in this book. Society is not in the same place as it was in the 1940s and 1950s. Back then, people wondered if a rocket could be built that could fly in space. Today, young children use the Internet to look at images of our Earth glowing in the Martian night sky. Every week, the Kepler telescope in space finds more and more Earth-like planets that could harbor life. And humans are making plans to establish colonies on the Moon and Mars.

PROGRESS TOWARD REICH'S SCIENTIFIC VIEWS

However tragic were the circumstances of Reich's death, he did succeed in getting the concept of orgone (or Life) energy into human consciousness in a new, scientific way. And the thrust of the development of science since Reich's death has been in his direction. This is not simply my opinion; there is evidence to support this statement.

In 1987, in his best-selling book, *The Turning Point*, physicist and systems theorist Fritjof Capra wrote of Reich, "...Wilhelm Reich was a pioneer of the paradigm shift. He has brilliant ideas, a cosmic perspective, and a holistic and dynamic worldview that far surpassed the science of his time and was not appreciated by his contemporaries."

In 2000, on a private visit with me to The Wilhelm Reich Museum, Harvard-Smithsonian astrophysicist Dr. Rudy Schild, recognized for his contributions to the understanding of gravitational lensing, dark matter, and black holes, immediately saw the significance of Reich's essential orgone energy experiment on display at the museum, the "To-T" temperature difference experiment. "If this experiment is valid, and I think that it is," Dr. Schild said, "then my science must be fundamentally re-evaluated."

In recent astrophysical discoveries, Dr. Schild and others have learned that there is measurable energy in space and that it has a "helical" or "spinning wave" movement as Reich described in the late 1940s and early 1950s. Dr. Schild references and credits Reich with the discovery of the "spinning wave" or *Kriesellwellen* form of the movement of the cosmic energy and with being the first to describe it based on scientific experimentation.

In 2003, Apollo 14 Astronaut and sixth man to walk on the moon, the late Dr. Edgar Mitchell, visited the Wilhelm Reich Museum with Dr. Schild and me. He, too, saw the importance of the To-T orgone energy experiment. He was especially interested in the Orgone Energy Field Meter; the Vacor Tube experiment; and the Cloudbuster. Dr. Mitchell observed, "Through these arrangements of materials, Reich is getting coherence out of the energy, and that accounts for the effects he is achieving." Dr. Mitchell was asked to describe what he meant by this statement. "I would use the analogy of natural daylight," he replied. "Natural light is incoherent. But laser technology makes the light coherent. The energy field of the planet is incoherent, but Reich's technology makes the energy coherent."

Earlier, in 1997, I had asked Dr. Mitchell what he saw in space out of the window of his space capsule with his naked eyes. In addition to the stars and other celestial objects, he said: "I saw the primordial units of energy in space with the naked eye." This phenomenon was completely unexpected, but it was seen by all the astronauts who left the Earth's atmosphere and entered the energy continuum of space. Dr. Mitchell saw what Reich's scientific experiments demonstrated—the mass-free primordial orgone energy units in the energy continuum of space.

A new generation of scientists may be better able to

understand what Reich is reporting in *Contact With Space* because of new developments and discoveries in science—especially in astrophysics and space science--over the last 60 years.

ON ENCOUNTERING NEW KNOWLEDGE

Reich often wrote that "everyone is right in some way." The challenge, he said, is to discover in what way. In *Character Analysis*, he observed that there is nothing a human being can experience that is not somehow real, no matter how "impossible" it seems. Again, the challenge is to discover in what way are such things real.

In addition, he stressed that the way of thinking is far more important than the facts discovered. The mechanistic-metaphysical way of thinking that dominates our culture will result in a different *interpretation* of the facts than will the functional way of thinking Reich used. These factors are key aspects of having an "open mind."

And an open mind is crucial to being able to hear what Reich is saying in *Contact with Space*. Letting go of the preconceived beliefs and unexamined assumptions we all have, that are part of the mechanistic-metaphysical thinking ingrained in all of us, is an essential prerequisite for the comprehension of the new knowledge offered in this book, knowledge which arises from a functional approach to Nature.

When the reader is able to enter into the way of thinking Reich employed and understand the conclusions he has reached, it becomes clear that his assertion— "Anything is possible. Nothing can any longer be considered impossible"—is actually true.

The terms "possible" and "impossible" are limitations defined by the present mechanistic-mystical way of thinking, not by Nature. The physics and technology of the 20th and 21st centuries are "impossible" according to the science of the 19th Century.

Giordano Bruno's understanding, more than 430 years ago, of an infinite universe with planets and life abundant throughout creation, is just now coming to be recognized by contemporary science. Kepler's Laws of Planetary Motion and Newton's Celestial Mechanics, from four centuries ago, are now used to guide spacecraft that are exploring our solar system. Quantum science began 100 years ago and was never accepted by Einstein. It took 80 years to begin to confirm some of its fundamental hypotheses (e.g., nonlocality and entanglement) experimentally. But today, the economy of the world depends on technology based on quantum science.

Six decades after his death, the jury is still out concerning Reich's science. A new chapter in the scientific investigation of his work is ready to be written by a new generation. It is my hope that you will read this thought-provoking book and come to your own educated conclusions. Most of all, as you embark on this journey with Reich, open your mind and heart and enjoy your exploration of the Cosmic Age at hand.

Mr. Mannion has been a professional writer and editor for over 40 years focusing on medicine, healing and new science. He has published articles, given lectures, and produced conferences about Wilhelm Reich and orgonomy. In addition, his understanding of orgonomy has been shaped by the direct experience of orgone energy through use of the orgone energy accumulator since 1971 and in orgone therapy.

WILHELM REICH — 1950
Bust by Jo Jenks

ELSWORTH F. BAKER, M.D.
Orgonomic Medicine, Financial Committee
1954-55

ROBERT A. McCULLOUGH
Research Associate—Oranur Chemistry
Expedition Operator, 1953-55

WILLIAM MOISE
Expedition Operator, Treasurer
1952-56

ERNEST PETER REICH
Student Operator
1954-55

STAFF OF OROP DESERT Ea, 1952–1956

EVA REICH, M.D.
OROP Desert Ea Physician
Expedition Operator, 1952-56

THOMAS ROSS
Operator-Caretaker, Orgonon
1952-56

MICHAEL SILVERT, M.D.
Operator New York, Transfer of ORUR
Treasurer, 1954-56

WILLIAM STEIG
DOR Emergency
Financial Committee, 1954-56

The contributions of the staff of the Expedition OROP Desert Ea were great in loyalty, exertion, sacrifice of time and work and risk of danger. These contributions cannot ever be forgotten.

These photographs are being published for the historical factual record of the OROP Desert Ea organization without regard to scope, extent, duration of service or lasting accomplishments of the participants in the U. S. A. desert expedition.

WILHELM REICH

CONTENTS

	PAGE
THOUGHTS OF IMPORT	XXVI
PREFACE	XXVII

There are no authorities—"What do they want for proof?"—Quest for knowledge governed by rules of learning only—Space war possible

I. TWO "STARS" FADE OUT 1

Does the Planet Earth Harbor Spacemen? 1

Realistic films on outer space—May 12th, 1954, 21:40-22:45 hrs., first contact with Ea (UFO)

New Tools of Knowledge Needed 4

Earthmen encounter space unprepared—First steps — Testing S.W.-U.S. desert—AAF jets helpless—Pilots ridiculed—Decision to complete independence of public opinion—Reliance on unbiased observation—Qualities first

Differentiation Between Ea and Stars 12

Astrophysics from scratch — Misleading names of constellations — Star tracks — One deviation

Two Successive Ea Photographs 19

		PAGE
II.	**THE "SPACEGUN"**	23
	The Atmospheric "ORUR" Effect	23

Preparations for the Arizona expedition — Changes in Oranur radium — Oranur 1951 and 1956

	Technological Use of ORUR	28

ORUR does not tolerate metallic restraint

	We Were Being Watched . . . ?	31

By U.S.A.F.—By atomic industry—by U.S.S.R. agents—By Ea

	Continued Contact with Ea at Orgonon, September-October, 1954	33

Spacegun = Cloudbuster plus ORUR—Ea moves against planetary pathways under ORUR influence—A second Ea is affected in its direction — Understanding the ORUR effects

	Conclusions Regarding Influencing Ea with ORUR-Spacegun	43

A motor force

	Summary of Available Data, October, 1954	45
III.	**CONTACT WITH THE U.S. AIR FORCE**	48
	Introduction	48
	First Contact with the USAF	48

Two Ea at Orgonon, January 28, 1954— USAF Technical Information Sheet, filled out

  PAGE

Survey on Ea 67

Report to AAF, March 17, 1954 (Injunction March 19, 1954)—Massfree gravitational equations—Integration of Kepler-Planck - $\mathcal{E}\,\sigma \neq \lambda^3 s^2$ — OR energy used by space invaders—Melanor (DOR) result of their using Cosmic Life Energy—Space not empty—Basic forms of cosmic space equations (notarized 1948)—Basic transformation of classical into orgonometric calculations—The kr^x system—Galactic propositions—The *"Swing"*—The *"Cosmic Pendulum"*

First Direct Contact with the Air Force Technical Intelligence Center (ATIC) in Dayton, Ohio 78

Conference on Ea—Report on disabled Ea

Technical Research Coordination 88

First major rainmaking operation Elsworth (1953)—AAF carefully follows Ea operations—Behavior of vapor trails—Behavior of clouds in DOR field—DOR clouds and high CPM—Preparations for possible Ea war—Negative gravity

The "Swing," ⌒ 95

The forward-moving, swinging pendulum—Characteristics of $\mathcal{E}\,\sigma$ — Expansion and contraction in the swing—Wave and pulse function—Form of the electrocardiogram—$\mathcal{E}\,\sigma \neq \lambda^3 s^2$

$l \neq 100\ t^2$ 101

	PAGE
IV. DOR CLOUDS OVER THE U.S.A.	111

    The Trip from Rangeley, Maine, to Tucson, Arizona, October 18 to October 29, 1954—3216 Miles 111

        Characteristics of DOR blankets—Desert DOR approaching

    Survey of Travel Route 123

    Transportation of the Cloudbuster and Laboratory Equipment 124

V. CHOICE OF TUCSON, ARIZONA 132

        Location in relation to Pacific Ocean—Galactic stream—Complete desert—Organization of expedition—Financial Committee—Equipment—Assistance

VI. THE PLANETARY VALLEY FORGE 138

    An Ea Attack on October 13, 1954 138
        The McCullough report

    The Tucson Base 143

        "Little Orgonon"—Position with regard to mountain ranges and galactic stream—First impressions—DOR "eats" mountains—Desert as a process—Erosion—Gullies—Remaining evergreen on mountain-tops above DOR ceiling

	PAGE
Desert Development	148

Killer DOR—Life on razor's edge—Energy equilibrium between two extremes of silence—Is DOR process reversible?—Thirst—Behavior of white Orene—"Black Rocks"

Phases 156

DOR attack upon green lands—DOR attack upon caked soil—DOR attack upon red clay—DOR attack upon the clay and the remnants of life

Proto-Vegetation—The Greening of Sandy Desert 158

Basis for greening of desert: *no rain*

Ea and Rain Clouds 161

Ea at Mt. Catalina—Ea and cloud formation—The greening experimental area

VII. THE Ea BATTLE OF TUCSON 165

Ea, DOR and Cloud Formation 165

Theoretical work positions November 10, 1954—Desert functions of DOR—Gradual adjustment of proto-vegetation and soil to moisture—Conquest of desert through proto-vegetation—AAF coordinated experimentation

[XXI]

PAGE

Establishment of an OR Potential on Mount Lemmon 172

Necessity for higher draw potential—Project to draw moisture from Pacific—"Ceiling" and "Pocket" DOR

"The Right to be Wrong" 174

Spaceship or Saturn? A photograph unexplained — Position with respect to Venus

The Breakdown of a Spacegun Operator 179

Paralysis while drawing from Ea

The Arrival of ORUR in Tucson, December 14, 1954 183

Instructions for transport of ORUR material—Effects—Silvert report

ORUR—Highly Excited 197

The Ea Battle of Tucson, December 14, 1954—16:30 Hours 199

The Nodes of Ea 201

Discovery of the energetic nodes of the Ea-CM (CORE Machines)

First ORUR Operations in the Desert 202

The January Oranur Rains 204

Gentle Oranur rains over deserts

	PAGE
VIII. BREAKING THE BARRIER	212

Search for Atmospheric Self-Regulation and the Obstacle in the Way 212

Greening of desert land should rule itself —Finding the western desert barrier

Experimental Quest for the Ea Barrier 217

High CPM Indicate "Atmospheric Fever" 217

High counts February 15, 1955—Meaning of high counts—The power of ORUR— The western barrier found

Probing the Atmosphere on Mountain Ridges 233

ORUR operation March 1, 1955

On the Way to the West Coast 238

The Barrier Yields 243

Rain over the American Sahara Desert

IX. COSMIC SELF-REGULATION IN OR ENERGY METABOLISM 254

Relationship of DOR and rain—OR and rain—OR and DOR—The dust devil— The DOR blanket absorbs the moisture— The DOR blanket breaks⟶ high counts —OR expels DOR—The dust storm—The DOR-Water-OR equation

LIST OF ILLUSTRATIONS

Wilhelm Reich, 1955 v

Wilhelm Reich, 1950—Bust by Jo Jenks vii

The Staff on OROP Desert Ea .... viii, ix

FIGURE

1. Star Tracks 17
2. Deviation of the "longer" track 18
3. Two Ea photographs 20, 21
4. Spacegun and operators 32
5. Ea moves under ORUR influence 35
6. A second Ea is affected 36
7. OR energy photograph and charts 39, 40
8. Birth and death of orgone units 42
9. Melanor on Observatory rocks 47
10. Ea in front of Spotted Mountain 65
11. Ea moves under ORUR influence 84
12. The "Swing," ⌒ 96
13. The forward-moving swinging pendulum 97
14. The energy of the swing, ⌒ 99
15. Basic form: "Electrocardiogram" 101
16. Map of travel routes 113
17. Mountain peaks clear above DOR blanket 116

Steig photo from Look Magazine.

FIGURE

18. Typical "Turret" shape in U.S. desert............. 119
19. East and west distribution of DOR................ 120
20. Position of the Tucson base 143, 144
21. Proto-Vegetation 160
22. Ea over Mt. Catalina 163
23. The greening experimental area at Tucson 164
24. Ceiling and pocket DOR.......................... 173
25. Position of UFO with respect to Venus, November 28th and 29th, 1954.. 175
26. Spaceship or Saturn? 176
27. What does this photograph depict?............... 177
28. Position of cigar shape relative to Venus between December 1st and 17th, 1954................. 178
29. Map of rainfall in U.S.A., December 10th and 11th, 1954 184
30. Ea-CM nodes 201
31. Ea-CM path 204
32. The base of operations........................... 216
33. Rainfall figures, February 16th to 18th, 1955 .... 227, 228
34. The "Barrier" at the southern Sierra Nevada 242

THOUGHTS OF IMPORT   (1953)

Thoughts of import
Are built like cathedrals,
Reaching high into the sky
As if to fly.

Onward they urge
From the depth of the brine
Pregnant with surge
Of ever greater design.

Let's burst open the sky
Let's reach for the stars
Let's ring out the cry
Transcending all bars.

Wilhelm Reich

PREFACE

There Are No Authorities

"What constitutes proof? Does a UFO have to land at the River Entrance to the Pentagon, near the Joint Chiefs of Staff offices? Or is it proof when a ground radar station detects a UFO, sends a jet to intercept it, the jet pilot sees it, and locks on with his radar, only to have the UFO streak away at a phenomenal speed? Is it proof when a jet pilot fires at a UFO and sticks to his story even under the threat of court-martial? Does this constitute proof?

The at times hotly debated answer to this question may be the answer to the question, 'Do the UFO's really exist?'

I'll give you the facts—all of the facts—you decide."

> Page 8, E. J. Ruppelt, "The Report on Unidentified Flying Objects," Doubleday & Co., Garden City, N. Y., 1956.

"What Do They Want for Proof?"

There is no proof. There are no authorities whatever. No President, Academy, Court of Law, Congress or Senate on this earth has the knowledge or power to decide what will be the knowledge of tomorrow. There is no use in trying to prove something that is unknown to somebody who is ignorant of the unknown, or fearful of its threatening power. Only the good, old rules of learning will eventually bring about understanding of what has invaded our earthly existence. Let those who are ignorant of the ways of learning stand aside, while those who know what learning is, blaze the trail into the unknown.

Quest for knowledge is Supreme Human Activity.

Nothing but the rules of *learning* can or should ever govern it.

April 1956.   *Wilhelm Reich*

The Case for Interplanetary "War"

(Quoted from C.R.I.F.O., ORBIT, Volume II, No. 8, November 4, 1955)

"Space War Possible is MacArthur Hint

Supporting CRIFO's beliefs are the timely and sobering words voiced by General Douglas MacArthur, before visiting Mayor of Naples, Achille Lauro on October 7, 1955 at the Waldorf-Astoria, New York. The Mayor revealed the General's statements to the New York Times as follows: '* * * He thinks that another war would be double suicide and that there is enough sense on both sides of the Iron Curtain to avoid it * * *. He believes that because of the developments of science all countries on earth will have to unite to survive and to make a common front against attack by people from other planets * * *.' The Mayor added that in the General's opinion the politics of the future will be cosmic or interplanetary."

This report contributes a few affirmative experiments to the opinion of a military leader quoted above.

Ea is the abbreviation for visitors from outer space, encountered in our globe's atmosphere, observed and experimentally tackled with the so-called *"Spacegun"* in various parts of the U. S. A. during 1954 and 1955 and the first half of 1956. "E" stands for "Energy", "a" for alpha or *primordial.* Ea was at times also referred to as representing *"Enigma."*

I suggest to establish a *Planetary Professional Citizens Committee* with sufficient work democratic organizational and statutory power to take over the responsibility for and direction of the Social Reconstruction of the Planet Earth and of the Ea Operations of the Future.

CHAPTER I

TWO "STARS" FADE OUT

Does the Planet Earth Harbor Spacemen?

On March 20, 1956, 10 p.m., a thought of a very remote possibility entered my mind, which, I fear, will never leave me again: **Am I a Spaceman?** Do I belong to a new race on earth, bred by men from outer space in embraces with earth women? Are my children offspring of the *first interplanetary race?* Has the melting-pot of interplanetary society already been created on our planet, as the melting-pot of all earth nations was established in the U. S. A. 190 years ago? Or does this thought relate to things to come in the future? I request my right and privilege to have such thoughts and to ask such questions without being threatened to be jailed by any administrative agency of society.

Many matters of my existence have, with this question, fallen quickly into place, having been uncertainties only four days ago; the temptation to answer the above question in the positive is irresistible. However, I shall postpone final decision until the facts have spoken. In the meantime I shall proceed on the assumption: *It is not beyond actual possibilities that men from outer space have landed (or will in the future land) on earth and have begun to breed here for whatever reason they may have had.*

This idea is not as foreign to the human race as it may appear on first encounter. In 1951, I believe, Hollywood introduced a new film with the title **"The Day the Earth Stood Still."** In this film a spaceship landed on earth; a spaceman, looking much the same as an ordinary, but very intelligent earthman, met with earthpeople. He fell in love with an earthwoman, made friends with her little boy. He knew mathematics far better than the mathematical genius of the earth century. The earth people got panicky since he was able to stop all earth traffic by pressing a button. In the end the earthmen did to him what they always do in such cases; they had their military hunt him and shoot him down in the street. He was brought to his spaceship by a robot; a new type of medicine, unknown on earth, revived him.

The film was excellent. It tended to prepare the population for extraordinary events to come. It had the right, and not the wrong, ideas about the functioning of cosmic energy used in the propulsion of spaceships. It pictured the spaceman as being akin to earthmen, but different in his attitude to women, in his behavior with a small boy, etc. It conceded that insoluble problems of mathematics were easily solved with his knowledge of the Life Energy. There was no doubt that the Life Energy was meant: Lights went on in the spaceship when fingers moved across certain switches; lumination of vacuum tubes can actually be achieved by approaching strong bodily energy fields.

All through the film show I had the distinct impression that it was a bit of *my story* which was depicted there; even the actor's expression and looks reminded me and others of myself as I had appeared 15 to 20 years ago.

I did not at that time have the thought that I could actually be a spaceman's offspring. Without my intention, somehow a ball of history started rolling, putting me in the center of space problems: I made actual contact by way of the cloudbuster with luminous objects in the sky on May 12, 1954, between 9:40 and 10:45 pm.

As evening planets 1954 were recorded:

> Venus: January 30 to November 15
> Jupiter: January 1 to June 30
> Saturn: April 26 to November 5

Latitude of Boston: Setting times:*

		Venus	Jupiter
March	15	6:55 pm	1:10 am
April	1	7:37 pm	12:13 am
April	15	8:13 pm	11:25 pm
May	1	8:53 pm	10:35 pm
May	15	9:24 pm	9:53 pm
June	1	9:49 pm	9:02 pm
June	15	9:55 pm	8:20 pm

During this hour men on earth saw for the first time in the history of man and his science *two "Stars" to the west fade out several times* when cosmic energy was drawn from them. The shock of this experience was great enough not to repeat such action until October 10th, 1954. The reason for the hesitation was obviously the risk to precipitate an interplanetary war by such experimentation. The event was kept secret.

I am a modest, orderly, rational man. I tend to understate my stature rather than to state my full potential as a scientist. I hate mystical irrationalism; but I believe that even in the most flagrant irrationality there must be some rational truth; there is nothing in this world of man that is not true somehow, somewhere, even if distorted to the utmost; this I used to tell my students. However, I never permitted irresponsible mystification of such serious affairs as interplanetary contact to confuse our grave task.

The Ruppelt report on UFO's clearly reveals the helplessness of mechanistic method in coming to grips with the problems posed by the spacemen. The cosmic orgone energy which these living beings are using in their tech-

---

* According to report by Robert McCullough. Local *mean* time. Even allowing for correction to daylight time, neither Venus nor Jupiter was in the observed area at the time. The objects were high above the horizon.

nology is beyond the grasp of mechanistic science since cosmic laws of functioning are not mechanical but what I term *"functional."* Even the mathematics necessary to formulate these problems and make them technologically usable, cannot use any of the old mechanistic methods of thought to cope with the functional OR facts. (See Orgone Energy Bulletin, 1950-1953.)

The helplessness of mechanistic thinking appears in the tragic shortcoming of our fastest jet fighter planes to make and hold contact with UFOs. Being unavoidably outdistanced is not a flattering situation for military pride. The conclusion seems correct: Mechanistic methods of locomotion must be counted out in coping with the spaceship problem. Neither propeller nor jet will or can ever get us into space beyond.

Easy contact was made on that fateful day with what obviously turned out to be a heretofore unknown type of UFO. I had hesitated for weeks to turn my cloudbuster pipes toward a "star," as if I had known that some of the blinking lights hanging in the sky were not planets or fixed stars but SPACE machines. With the fading out of the two "stars," the cloudbuster had suddenly changed into a SPACEGUN. From then onward, too, our approach to the problem of space became positive, affirmative, confident in using our carefully screened data.

New Tools of Knowledge Needed

When I saw the "Star" to the west fade out four times in succession, what had been left of the old world of human knowledge after the discovery of the OR energy 1936-1940 tumbled beyond retrieve. From now on everything, anything was possible. Nothing could any longer be considered "impossible." I had directed drawpipes, connected with the deep well, toward an ordinary star, and the star had faded out four times. There was no mistake about it. Three more people had seen it. There was only one conclusion: *The thing we had drawn from was not a star. It was something else; a "UFO."*

I must remind the reader that in May 1954 I had read only one report on UFOs; I had not studied anything on the subject. I knew practically nothing about it. But my mind, used to expecting surprises in natural research, was open to meet anything that seemed real. I had to be convinced myself first. Most people try to obtain consent of their impressions before having been convinced themselves. I had long since given up hope to convince anyone steeped in present-day mechanics or mysticism. There were no authorities. There was no one to whom to report this observation.

(In May 1954, the assault by the American drug business had just begun to bother us a few weeks before.) We were still laboring at an understanding of what had happened in early 1951 when Oranur had burst into our lives; we were still trying to dig out humanly, emotionally, and scientifically from under the avalanche of new observations, facts, ideas which Oranur had thrown in our way. I knew, we had without intending it, drilled a hole, as it were, into the wall which had for millennia separated man from the universe around him. We were hard pressed in our attempt to survive the flood of events in good form. A U. S. court of law had issued an order on March 19th, 1954 to stop all OR research activities including publication. We thus had to face the flood of incredible new facts, our own emotional and physical misery and the assault by the American and Russian mechanistic mind. It all tied in with our basic research neatly as one single fact: **Earthmen had encountered space as it really was**; not as science had conceived of it heretofore.

Here is the sequence of the main events in OR research which had preceded the fade-out of a "star":

On December 30th, 1953, one of our workers reported a "sighting," an orange flying object traveling from SW to NE.

On January 9th, 1954, I saw the Wells film "War Between The Worlds," a rather realistic approach to the planetary emergency. The film still adhered to the earthly idea that war consisted in a noisy bang-bang of shooting machinery; that war upon the earth waged by beings from outer space would produce the sight of green arms, clawed hands, and other products of human nightmare phantasies. The film was ignorant of the insidious, noiseless, emaciating nature of a war of attrition: *Drawing off life energy from a planet covered with green, seething with living beings.* However, the presentation of landings was realistic, and so was the crumbling of buildings. We may be right in admitting that the crumbling was filmed with high speed acceleration of an actually nearly invisible activity. The panic shown in people was due rather to the acceleration of events than to the events themselves. In reality, the rocks were already crumbling in 1952 in the U. S. A., but no one seemed to pay attention to it, since the process was so very slow and the people who noticed it were so very evasive or fearful to tell.

During all of 1954 preparations for our expedition to Arizona and other parts of the southwestern U. S. A. were under way, greatly handicapped financially by the assault of the food and drug industry upon our financial resources, the income from the medical use of the OR energy.

During the summer of 1953 I had initiated the pre-atomic chemical research project in order to obtain, if possible, an answer to the problem of Melanor, the black powderlike substance that came down at Orgonon from the atmosphere. It attacked and destroyed rocks, dried up the atmosphere, made us miserable with thirst, cyanosis, nausea, pains of all sorts.

In April 1954, on the basis of our findings, Mr. McCullough went to Yuma, Arizona, for 4 weeks with a cloudbuster to find out about DOR clouds and Melanor fully developed in the desert. The results were rewarding and will be dealt with in connection with pre-atomic chemistry problems of outer space. It was McCullough who first reported on the *"DOR ceiling"* in the desert, thus corroborating the finding of stagnant DOR clouds over Orgonon.

These preparations were replete with observations of an extraordinary nature. They led up to the startling event of May 12th, 1954: *the fading out of a star*. They also sharpened our senses and prepared us for the formation of new methods to observe Ea correctly. It was towards the end of 1953 that I read Keyhoe's report on unidentified flying objects.* It struck me at this first reading on the subject that no one had as yet tried simply to make a series of photographs of the night sky. This is at least the impression one gets from all published reports, including the one by Ruppelt (1956). The reports are filled with eyewitness reports on details of location, time, direction and similar things. However, no one, as far as I can judge, had started a systematic scanning of the night sky. The answer seemed to be: No one knew *what* to photograph and *when*.

Another striking feature was the frustrating effort of our jet pilots to outrun what quite obviously were technologically far superior craft. I had at times the impression that competitive ambition and social anxiety of "authorities" rather than sober, systematic approach ruled the efforts to get at the UFO problems. This made it easier for the neurotic distracters from the serious UFO problem to deny the whole affair as an impossibility and to denounce those who claimed to have observed extraordinary matters as cranks, lunatics, swindlers or worse. We understand how the serious pilot who was outdistanced

---

* Donald F. Keyhoe—Flying Saucers from Outer Space, 1953, Henry Holt.

by the UFO must have felt jammed in between ridicule and hurt pride. All this did not help, of course.

Seen from this point of view, the attack by the food and drug agents upon orgonomy was no more than an expression of terror of the Cosmic Energy on the part of sick men. It, therefore, *"did not exist"*; its discoverer was to this view either a "quack" or a "lunatic."

It may well be that these first impressions were responsible for the development of systematic techniques of observation first of all. Quantitative, exact measurements could without harm wait until one could get hold of some reliable, distinguishing observations of the sky.

I therefore encouraged my assistants to make no notes, to take no photographs, to make no attempts at a quantitative determination of what they thought were worthwhile observations; but rather to relax and just look into the sky, casually, as it were, and let the new facts come to them rather than to run breathlessly after elusive, strange facts.

This idea paid off in the long run. We learned gradually to find out the UFOs among the myriads of stars and to test our judgment by way of the Spacegun and the camera. From this kind of approach resulted the positive findings reported here. We were always careful to let new observations *force* us to pay attention to them. Observations should be rejected until they return again and again and are no longer deniable. Questionable observations which did not return again and again were to be neglected. We refused, too, to drown in paper protocols of heaps of uncomprehended events. We had to obtain a firm foothold first. Before we would measure we would try to obtain the *qualities* of what should later be exactly measured. I felt that, in order to please the "authorities," the observers overloaded their reports with unnecessary, even misleading details. What was important was first of all *not where* a UFO was seen or *when,* but what impression it made, how it affected the observer both emotionally and physically; not whether the observer was a pilot—pilots can be irrational, as anyone else—but whether he talked straight; not the exact altitude of the UFO, unimportant at this point, but whether its movements were out of the ordinary, unknown. And last but not least, in order to obtain reliable results, regardless of whether positive or negative, one had to assume that the object under investigation was real and worth the effort. It is senseless to eat a sandwich and to proclaim at the same time that it does not exist. If it does not exist, then leave it alone, don't eat it, and keep silent, to say the least.

Here we have a few rules of conduct to follow in the study of UFOs. These rules, far from complete, emerged from the process of observing the sky during the past few years. These rules have been corroborated by the results obtained in Desert Research and practical work with DOR and rain clouds.

Having become acquainted with the rules of observation we shall gain confidence in our judgments and in the opinions of others. We shall no longer hang on to the tails of public opinion or to a non-existent authority on matters utterly unknown and strange. We shall gradually become experts ourselves in the mastery of the *"Knowledge of the Future."*

Having freed ourselves from our dependence both on public opinion and on our own urge to please, impress or convince non-existent "authorities"; having learned how to rely on our observations only, we may proceed with confidence in our observations and experiments.

First, we make ourselves at home in some region of the globe and of the sky by living, as it were, continuously with our field of operation. I knew the region around Orgonon well within an approximate radius of 100 miles. *Daily* routine trips through this region, equipped with field glasses, Geiger counter, photographic equipment and other necessary instruments made it possible for me to see nuances of change in vegetation, changes over only a few weeks or even days, the role mountain ranges play in the drift of clouds; the shape of clouds over certain areas; the "feel" of dipping in and out of what later came to be called "DOR pockets," etc. It also made it possible to adapt myself to the particular changes in the *background counts* which characterized the region. For example my report on the blackening rocks (OEB, 1952),* tells of a wedge of high counts into a region with near normal background activity on routes 17 and 2 in Maine near Rumford. Without the knowledge of the shape of the mountain ranges in relation to the location of Orgonon such a finding may have escaped me.

During the Arizona Ea operations, 1954/1955, as well as during the Washington Ea operations 1955/1956 a certain region was designated as research territory. The rule was strictly adhered to, to drive a certain route daily from and to base, 80-100 miles in Arizona, 40-50 in Maryland, to get acquainted with the region, to know its vegetation well, even single trees or groups of chollas, certain patches of desert sand, certain dried-out river beds on mountain ranges, the *bottoms* of river beds in the valleys, etc. We learned during these operations to avoid any pretense of exactness. Taking pictures immediately or measuring something without knowing what one measured, why or to what end, or writing endless protocols, was only confusing, empty of true knowledge. The rule was strictly adhered to to know the research object well by mere observation and coordination of observation, *before* quantitative tests were made. This secured crucial coordinations

---

* *Orgone Energy Bulletin.*

such as background counts with cloudforms in the sky or DOR accumulation over the region. It led to important conclusions regarding the effects of Ea on climate and vegetation.

There is little use in determining quantitative properties of an unknown object or realm before having obtained sufficient orientation about its basic qualities, appearance, behavior, functioning, inner dynamics, development, etc. It not only tells nothing whatever about the *unknown;* it prevents us from really getting at it. It gives us the illusion that we know something if we can determine "exactly" at what hour, minute, and second, at what latitude and longitude, etc., we saw the unknown. This is important where *locations* of *known* objects are concerned. It is of secondary or of no importance, it may even distract from the issue at stake where unknowns are concerned. There is no use wasting time and deceiving oneself about being an "exact scientist" in counting the number of words a man talks per minute at a certain hour of the day in a certain building, if, what bothers him and drove him to us as doctors, is a severe bellyache which could not be diagnosed by scores of physicians. The first thing to do in such cases is to relax, sit down, look at the patient, chat with him and get "the feel" of him, a feel so very much despised by "exact scientists" as "psychology" or "mysticism." Such "scientists" are no scientists at all. They are ignorant with respect to the crudest, basic scientific rules of observation; to the intricate relations between observer and observed; to the inevitable influence of sense impression and emotional structure of the observer on the observed. A compulsive-neurotic doubter in science will doubt and talk away everything, no matter how much evidence there is. An arrogant stiff-neck in science will remain inaccessible to any kind or amount of most impressive demonstration. It does not change the situation a bit that the sick man hides his compulsion, ignorance or arrogance behind a veneer of fake scientific objectivity. Being "scientific" means being open-minded, ready to accept anything if certain conditions of objectivity are fulfilled; of being *positive,* and not negative in examining a new realm of knowledge. These are banalities indeed, as old as good, true science itself.

I shall present on the following pages some drawings and photographs of phenomena unknown to science. I do not present them in order to convince anyone. The sole objective is to explain some of the methods used in the Ea work. I shall omit dates, locations and similar data in order to concentrate on the *principle* of reliable observation, and not on secondary matters of presentation of facts irrelevant at the moment. This will also help in keeping the frightened, neurotic, "critical scientist" off the scene of our serious debate.

Synoptically the following characteristics of unknown objects, Ea in our terminology, were found important in celestial observation:

Differentiation Between Ea and Stars

Known

Unknown, Ea, UFO

1) *Color* of fixed stars: steel blue; flimmering on clear days; not flimmering before onset of rain.

   Color: Yellow to red or white, flashing, pulsating. Changing color from yellow to red or green and vice versa.

2) *Size* of fixed stars: smaller than Jupiter.

   Size: Mostly larger than Jupiter.

3) *Location* in known spot, nightly visible according to known astronomical schedule.

   Location irregular upon systematic nightly observation, Ea *sometimes absent,* or appearing suddenly in middle of sky, *vanishing suddenly,* etc.

Known	*Unknown, Ea, UFO*
4) *Movement* lawful, predictable.	*Movement* at times regular in accordance with either fixed stars or planets on ecliptic. But also *deviating grossly* in *speed and direction*. Photographic relation to fixed stars important. (See deviation, pp. 17, 18.)
5) *Known flying objects* like planes move at known speeds; straight or curved; balloons drift with the wind steadily, never against the wind.	An Ea may appear like a star on Eastern horizon, then move slowly *along horizon* to south and sink below horizon in west. Ea of the silvery type change speed abruptly, go against the wind, wobble like spinning tops, disappear suddenly, affect electro-magnetic instruments, *cause clouds to disperse.*
	Ea hover in sky, on mountain slopes; maneuver; give impression of watching ground beneath. Often move with stars in setting; but also may follow no rules at all.
6) *Spacegun effect* dimming, no fading. Only general region darkens somewhat. No special sensations.	Upon correct drawing off energy there follows *slowing* down, *dimming, fading, disappearance, change of direction.*
	Sour or bitter taste on tongue, nausea, at times loss of equilibrium.

Known	Unknown, Ea, UFO
7) *Meteorites* streak across a stretch of sky with a bright flash, beginning and ending sharply, occurring in *irregular, varied* directions, mostly toward the ground.	May appear to be meteors. Ea only visible with strong field-glasses upon prolonged careful observation of certain stretches of sky in very high altitudes, moving either in groups *synchronously* or in *quick succession* over the *same area* in the *same direction*, with approximately the *same speed* as if in military formation, often parallel, not perpendicular to the ground.

Ea is a new event without precedent in our lives. Humanity, with the exception of a few philosophers, had no idea of the possibility of visitors from outer space. *Earthman* had not developed any view, method or scientific tool to cope with the problem. In addition he has developed in his offspring a character structure and a kind of thinking which obstructs the approach to the new fact by way of ridicule, slander and outright threat to the existence of the pioneer of space engineering. Therefore, our new approach must *start from scratch, as if* no science existed at all. Men, upon return from the moon, when asked whether there are barber shops on the moon similar to that on the corner of 42nd St. and Broadway in New York would say: There are none; period. Neither are there "authorities" on the moon, we must add. There are no authorities on these new matters.

Astronomy has to start from scratch. And "scratch" here signifies the fact, that both the Copernican as well as the Keplerian theories, the "perfect" circle and the "perfect" ellipse, are inapplicable to the true movements of the solar system. The sun moves onward while the planets "circle." There are, accordingly, no closed circular or elliptical paths around the sun. The circles and the ellipses are necessarily replaced by *spinning waves* (KRW, Swings, ⌢) in various spatial relations to the sun, by irregular accelerating and decelerating speeds as in spinning tops, by curved general pathways, etc. This example may suffice to clarify the inevitable new approach to astrophysics.

It is, therefore, understandable that a panel of top scientists established in 1952 to screen the UFO came to the following conclusion:

"We as a group do not believe that it is impossible for some other celestial body to be inhabited by intelligent creatures. Nor is it impossible that these creatures could have reached such a state of development that they could visit the earth. However, there is nothing in all of the so-called *'flying saucer'* reports that we have read that would indicate that this is taking place." ("Report on Unidentified Flying Objects," 1956)

Ruppelt continues:

"The Tremonton Film (an actual film of UFO's) had been rejected as proof but the panel (of experts) did leave the door open a crack when they suggested that the Navy photo lab re-do their study. But the Navy lab never rechecked their report."

It is quite understandable that the U. S. Navy did this. One cannot judge functions of one element, while living in another. One must *live* things to judge them. The problem lay buried in the vast realm of the Life Energy.

Though the present theories are wrong, do not reflect space reality, the calculations are correct; the predictions of astronomy, if I am correctly informed, are true. I know nothing about the behind-the-scenes affairs within astronomic circles. However, there is little doubt as to the confusion created by the old Egyptian and Greek names given to *"constellations."* They are artificial, correspond to no reality. Man's ideas about his gods and not functional groupings of the stars, if any, determine the *"constellations."* They should be abandoned as quickly as possible, since they are misleading, preventing us from looking at the stars unbiased by arbitrary designations.

I confess that I never could remember well the various god or animal names; nor could I ever make out any similarity with actual star patterns. Other groupings seemed just as possible. I succeeded in the course of decades to lose my feeling of ignorance. I began to see most thrilling patterns of groups of fixed stars which suggested groupings according to superimposing spinning waves. I may be entirely wrong in this. Neither would I like to add to the existing confusion and paralysis of astrophysical thought. Instead, I would like to present an example of naive star photography. The following photograph was taken one night during the expedition in Arizona with no special objective in mind. I needed a photograph of pathways of stars. I am omitting concrete data since they are irrelevant here. They should not distract from the main subject.

The photograph of star-tracks was taken with a Leica, with aperture 3.5, at night without moon, directed toward a certain group of stars near zenith:

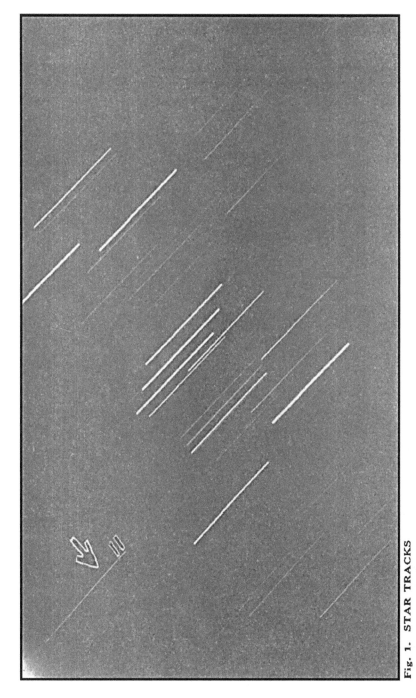

Fig. 1. STAR TRACKS

Exposure 20 minutes; full aperature 3.5; moonless night; *track with arrow is longer than all the others; parallel* lines; incidental finding: the longer track deviates by approximately 2.5 degrees.

At first sight the photo shows no irregularities. Upon careful study, however, *one track* (see arrow) proved on the enlarged photo about 2 millimeters *longer than all the other tracks, which are exactly identical*. This little piece of evidence is enough to justify the *"FROM SCRATCH"* point of view. The quantitative difference is irrelevant here. The *qualitative* fact of one track being longer is what matters.

Upon closer examination of the tracks we find a deviation of the same "star" track from the parallel run of the rest of the tracks by about 2.5 degrees. We shall strictly withhold any idea or opinion of what this astounding fact may indicate. We must proceed without submitting to being shackled by theory.

Fig. 2. Deviation of the "longer" track

Two Successive Ea Photographs

The following two photos were taken on two consecutive clear nights at Little Orgonon in Arizona during the winter months of 1954/1955. Again, exposure, exact hour, day of the month are here omitted as irrelevant. Only what we *see* counts.

These are extraordinary celestial phenomena. The time of night, the duration of exposure, the aperture and position were the same in both pictures. The identity of the location is evident from the identity of the landscape. The brightness reflects the lights at Tucson to the south of Little Orgonon, some 8 miles away. The two pictures differ greatly, contrary to astronomical pictures. They present tracks of Ea in terms of Orgonomy, UFOs in the parlance of the Air Force Technical Intelligence:

PHOTO A: An Ea under ORUR to the LEFT (East) of trees. Note disruptions—"fading out."

PHOTO B: Ea, appearing suddenly (*middle* arrow, to the right of trees the following night at the same time and same exposure). Two upper arrows mark *deviation* and three times "fading out."

Fig. 3. Two Ea photographs

1. Photo A shows the track low above the tree to the left (or east), while Photo B shows the same object to the right (or west) of the same tree.

2. The track in A is much shorter, nearly half of the track in B during the same period of exposure. Track B appears suddenly in the photograph as if the object had entered the sky from nowhere, as it were.

3. Track A shows several interruptions: The object was still there in the dark areas, but did not luminate.

4. Track B shows no such interruptions in the emission of light. **It appears SUDDENLY in the photograph** (middle arrow) as if from nowhere, moving westward (to the right). It appears suddenly while another object moves apparently much faster higher up. This second object appears on the photograph plate at the beginning of the exposure. It shows a "wobbling" curve at the first arrow and three regular interruptions of light emission at the third arrow.

5. There are on the Photo B a few other parallel curved lines fainter than the lower two. The Photo A does not show any of these additional tracks.

We shall forego any attempt at an interpretation. It is the set purpose of this account to train ourselves in suspending judgment, in omitting well available quantitative description, in not telling when these photos were obtained. We must keep this first approach to the Ea problem strictly *observational* in order to find a first orientation as to *what* to look at, what to expect as being **"unknown"** in the sky. We do not wish to be negative, obstructive. *We want to be affirmative in our approach;* affirmative in a careful, step-by-step, observational manner.

CHAPTER II
THE "SPACEGUN"
The Atmospheric "ORUR" Effect

The summer of 1954 was mainly spent with preparations for the expedition. We had learned in two parallel lines of observation that the white and the black powdery substances, Orite and Melanor, were related to drought, dryness and discomfort in man, animal and vegetation alike. At the same time, a relationship was established between these atmospheric conditions and the presence of the big, yellow and reddish pulsating "stars" which we had learned to distinguish from planets and bluish fixed stars. But, as is to be expected in such situations, we were not sure by any means that the established correlations between Ea and drought were correct. Therefore, the old idea to go to the fully developed desert and to test our observations there, gained in importance as the weeks passed by.

The preparations became concrete. With these preparations getting along well, it struck me as peculiar that I had so completely abandoned the original Oranur process of 1951. When the atmospheric DOR emergency had become unbearable at Orgonon, the small amounts of nuclear material, three milligrams of radium and a few micrograms of radioactive cobalt, had been buried in an uninhabited area on Route 17 toward Rumford, about 15 miles from Orgonon. For more than three years now this material had been buried in the ground within heavy lead shielding. Why not test this material again before leaving for Arizona, I thought. The idea seemed well-founded. On September 21, 1954, the radium was brought back to Orgonon and was tested immediately with the 4096 Tracerlab Auto-Scaler.

The results were recorded in the following original protocol. In order to compare with the original activity on April 28, 1951, the report of that date is reprinted here in full regarding Radium needles 1 and 2:

A. **Oranur Experiment, First Report, 1947-1951—pages 323-324**

Measurements of Two Needles of Oranur Radium (one mg. each), in Shielding and Naked, at GM Autoscaler, *April 28, 1951, 3 P.M.*

All measurements at one cm. distance. Each measurement average of several measurements.

Material	Scale	With Shielding	Naked	Time in sec.	CPM
1. Ra 1 (1 mg. untreated)	4096	+	....	0.8	307,200
2. Ra 2 (1 mg. OR-treated)	4096	+	....	1.05	245,760
3. Ra 2	256	....	+	0.4	43,000
4. Ra 2	4096	....	+	2.8	81,920
5. Ra 1	4096	....	+	8.3	28,877
6. Ra 1	4096	+	....	0.8	307,200
7. Ra 2	4096	+	....	0.8	307,200
8. Ra 2	4096	....	+	3.0	81,920
9. Shielding alone	64			3.15	1,280
10. Microgram Ra—OR-treated 5 yrs.	4096			0.8	307,200
11. Watch, one month owned	4096			10.0	24,576
12. Watch, 2 yrs. owned	4096			10.0	24,576
13. Calibration after measurements	256			4.15	60

Before proceeding further, let us again review the facts in their interrelations, and *not* **singly**:

FIRST: *The naked NR material gave a much lower count* (**one tenth**) *than the same material enclosed in heavy lead shielding.*

SECOND: *The ten times higher count in the atmosphere around the shielded NR material is a function of the OR energy fighting against NR.*

THIRD: *As soon as the interaction between OR and NR is stopped, the high OR activity vanishes and sinks down to the normal atmospheric level.*

FOURTH: *OR energy alone does not react severely unless irritated by NR.*

B. Oranur, Second Report

Protocol (upon arrival): *September 21, 1954*, 12:00 noon

Subject: G-M reactions to **Oranur**-affected radium

Present: Wilhelm Reich, M. D., Thomas Ross, Robert McCullough

At 12:00 noon today Mr. Ross brought back to the Observatory the box containing the 3 milligrams of radium,* which had been buried 15 miles to the southwest of Orgonon since March 31, 1952. An immediate background count of the general area was made with the SU-5 survey meter. It gave 150 cpm. With the unshielded counter tube held one meter away from the box of NR, the count was 8-900 cpm. With the tube laid on the top of the box, it was 8-9000 cpm.

One 1 mg. radium needle was then tested with the larger and more sensitive G-M tube and Autoscaler. With

---

* The *third* milligram was referred to in the First Oranur Report (1951) as the last control NR, on pp. 324-325.

the needle naked at a distance of 1 centimeter the count was 2457.6 cpm (against 16,000 in 1951, WR). With the radium needle reinserted into the lead case it had been enclosed in and the counter tube 1 centimeter distant from the case, the count was 163,840 cpm (against 7000 in 1951). Notes: One water-tight lead radium container was found to be almost full of what appeared to be water. The screw cap was greased and no plausible reason was evident for the presence of the water. WR had the box of NR placed touching the cables of the Cloudbuster. Immediately following this, a general feeling of well-being was marked in two persons. There was a sudden brightening of the rocks and vegetation; distant mountains, which had been black, suddenly became blue; the sky cleared and the west wind started. WR remarked that his organism seemed to be filling up. RMC noticed that his hands were filling out to their former fullness. Deep full breathing could again be enjoyed. WR's dog Troll showed no aversion to the box of NR.

/s/ Robert A. McCullough

C. **Oranur, Second Report**

Protocol: *September 26, 1954*—10:30 hrs.

Present: Wilhelm Reich, M. D., experimenting.
William Moise, Robert McCullough, taking protocol.

Place: Orgone Energy Observatory, Orgonon, Rangeley, Maine.

Subject: **G-M Rates of Oranur-affected NR** (of 1951).

/Signed/ Wilhelm Reich

The large double lead container containing one of the 1 mg. radium needles that had been used in the **Oranur** Experiment was brought into the Observatory and the C.P.M. rate was taken with the 2.3 mg./cm² counter tube and the Tracerlab Autoscaler. The results follow:

1. Radium *in* the double *lead container*—1 cm. distant
    Scale 4096   10.65 sec.   24,576 C.P.M.
                    11.65 sec.   22,000 C.P.M.

2. Radium in smaller, inner *lead container*— 1 cm. distant
    Scale 4096   2.50 sec.   98,304 C.P.M.
                    3.65 sec.   70,246 C.P.M.
                    2.00 sec.   122,880 C.P.M.
                    59.10 sec.   4,096 C.P.M.

3. Radium needle *naked*—1 cm. distant
    Scale 256   13.50 sec.   1,024 C.P.M.
                    11.42 sec.   1,400 C.P.M.
    Scale 4096   5.5 minutes   702 C.P.M.

4. Empty large lead container—at 1 cm. distant (*Radium removed*)
    Scale 64   1.0 sec.   nearly zero
                    1.0 sec.   nearly zero

5. Calibration after measurements
    Scale 256   4.15 sec.   60 cycles
    Threshold voltage 1200 volts.

/Signed/ Robert A. McCullough
Research Associate

Radium 1 and 2 had lost much of their activity of 1951. The third needle which had not been treated with concentrated OR energy was much lower than even its own original rate of 16,000 CPM naked at 1 cm. distance from the counter tube. The great vacillation, the activity of the material, is as striking as the *far higher rate within heavy lead shielding.*

These were basically new phenomena. The basic feature of a far stronger activity *within* shielding remained to this day. The activities *without* shielding varied greatly all through, since 1954, with place, climate, transportation, etc. It was heretofore impossible to find any mechanical lawfulness in these oscillations.

Technological Use of ORUR

The new use of the Oranur material in connection with the desert project in Arizona came about in the following manner, while we were searching desperately to alleviate our physical distress. I quote from my Log Book, date September 21st and 28th, 1954:

September 21.

McCullough packing truck * * *

My ankle swollen from bad DOR at Orgonon; left ankle hurts, swollen and inflamed * * * (my family gone * * *) I am alone, abandoned here.

September 28.

I called Moise and Eva to be at Orgonon around October 1, for packing * * * Packing proceeds.

Ea this morning 5:30 a.m. very strong till 6:30 a.m., drawing off energy, felt by WR (causing fast heart, 2 x vomiting, fatigue, depression); Helen Tropp woke;

Mrs. McCullough felt weak, nauseous; Mrs. Tom Ross, too * * * I felt like giving up, dying * * * A bright yellow star was hanging low in southern sky. Time to go to Arizona.

Oranur Weather Control richer by reinstatement of OR + NR, stopped by separation, with NR + Pb (container) 40,000 CPM.

DOR easily removed by OR ++++ (meaning OR + Pb).

Principle: Irritated OR causes Expansion. Sky clears, blue within minutes.

This was a tremendous step forward. Heretofore, the clearing of the atmosphere had been done by drawing off the DOR clouds into a lake. Now, within a few seconds, the sky cleared and became blue in zenith and far around the horizon upon **Orurization**. The name ORUR was later coined to distinguish this operation from **Oranur**. In the latter, *original* nuclear material (NU) irritated *concentrated* atmospheric OR energy, making it run amok or change to DOR. Now, on the other hand, NU, rendered harmless by Oranur and weakened in radioactivity, was changed to OR behavior in the opposite direction from **Oranur**, as *ORUR*. *It was low energetically when not within lead or metal of other kind. However, as soon as it was put into lead the counts climbed instantly outside the lead container up to dozens of feet to 40,000 or even 100,000 CPM measured with the SU-5 Survey meter.* This was news, indeed: **The atmosphere could be charged, "Orurized," directly by ORUR.**

I repeat: *The Oranur material gave naked no, or only a negligible, GM reaction. The lead container, too, gave no reaction. But the moment the two were brought together as far away as from two to three feet, the GM counter (SU-5) soared to 40,000-100,000 CPM.* This was astonishing as it was incomprehensible. It was unknown and somehow weird. A few days' observation revealed the fact that ORUR was an extremely powerful tool. If used only a few seconds, two to five, it cleared the sky of DOR. The DOR hovering low and dark over the landscape, especially in the valleys, seemed to turn blue-gray from horizon to horizon. If used too long, depending on the weather from 20-60 seconds, clouds began to form nearly instantly; and rain would set in a few hours later. Also, gray even clouds that "could not pour out their water," gave a gentle downpour immediately after Orurization. This was found out the hard way when we overdid once during the first few tests. Later I learned to respect the great sensitivity of the atmospheric energy and the power of Orur.

The reaction of the atmosphere seemed more effective when the Orur material was moved in and out of the lead container. At first, Orur was used apart from the cloudbuster. But soon the two were combined. We would draw off some DOR from zenith and around the horizon first and then *"orurize"* from the west or southwest. The southwest direction seemed particularly effective. We soon learned that touching the southwest-northeast galactic direction (see CORE, Vol. VI, 1954, p. 92) for a few seconds would make the blue-gray OR energy stream in rapidly, covering the formerly blackish dirty-looking mountains with a fine blue-gray haze. The change in the atmosphere was immediately felt by all observers. Even dirty steel-gray DOR-affected rain clouds seemed to fill up and become white in a brilliant, formerly dull, stale atmosphere.

The idea forced itself upon me: *It will be possible to draw off DOR from the deserts and draw in moisture from the Pacific Ocean.* The orurization effect of September 29th had reached as far as 170 miles toward the coast where my daughter lived. She had witnessed a sudden clearing of DOR and a fresh brilliance in the atmosphere. She had thought that I had done some new experiment to the west at Orgonon.

We Were Being Watched...?

New facts which are out of step with routine experience should not be denied or ignored; they should be pushed aside again and again until they force themselves upon the observer as true and objective and can no longer be ignored. This was done with the impression we all had had during the preparatory operations at Orgonon: *We were being watched.* It is still hard to believe it today. However, a symphony of relevant facts forced its way through to reality. There were three series of processes which were related to this. They were initially independent of each other; they were experienced separately without communication by three different groups:

1. We had begun "drawing" energy from Ea; their lumination was weakened or even extinguished; they moved conspicuously out of order on several occasions at Orgonon during 1954.

2. McCullough went through an experience of the strangest sort on October 13th, en route to Arizona, near Kansas City (see his letter of October 19th, 1954, p. 139). This event fell into line with an event we witnessed on December 14th, 1954, the day when Orur arrived by plane at Tucson, Arizona.

3. Moise was ordered to report our contact with Ea to the ATIC headquarters in Dayton, Ohio. The Air Force officers who received this report were burningly interested, but seemed not surprised at what he told them. (See verbatim report by Moise, p. 79).

We shall deal with each of these independent processes separately, but we shall keep them in our minds integrated as parts of one single process Ea:

Oranur operations on earth seem to be carefully watched by living beings from outer space.

Fig. 4. Spacegun and Operators:
E. Peter Reich
M. Silvert, E. Reich, W. Moise
OROP DESERT Ea, Washington, D. C., 1955-56

Continued Contact with Ea at Orgonon, September - October, 1954

On September 29, 1954, while orurizing the atmosphere on the hill near the Observatory, I saw, looking skyward above me at 10:45 a.m. four black-gray craft traverse the Observatory region at high speed, unusual for earth craft, including the fastest jets, at great altitude within two seconds from northeast to southwest. They flew in formation.

A potential of 40,000 to 60,000 CPM had been artificially created a short while before with ORUR. I orurized the atmosphere daily now since the DOR effects returned promptly a few hours after having been removed. The *"Spacegun,"* i.e., the integration of ORUR and Cloudbuster was conceived by me on the same day:

Protocol: September 29, 1954   **Invention of Spacegun**

Subject: **Combination of ORUR and Cloudbuster**

WR informed me today that he conceived the idea of the combining of **Oranur** and the Cloudbuster for the enhanced operation of both. WR further stated that he will try a long **Oranur** reaction with the next hurricane that develops.

/Signed/   Robert A. McCullough

The technological development of Oranur toward ORUR operations rushed onward speedily in logical steps.

The morning of September 31st was striking. We had cleared the sky. An extreme, fast-forming humidity deposited heavy moisture on the window panes, on cars, on rocks and on meadows. The temperature and climate felt subtropic. Brown and yellow leaves were turning green as we looked at them, observed by McCullough, Tom Ross, an old woodsman, and myself, raised on a farm. All weather reports were upset that day. The Air Force seemed very active overhead; many planes were circling our region. A red plane flew five times over and around Orgonon. We were used by then to these inspections of our grounds from the air by various airborne observers. We had cleared the sky. Our feet felt heavy when we walked the steps to the Observatory. This phenomenon was well known to us as a consequence of strong Oranur; the pull of gravity was stronger (to be dealt with in a different context).

On the evening of October 5th-6th four of us saw three large, pulsating, yellow Ea hanging low over the southern horizon and one to the northwest over the Observatory. They had not been there the night before.

McCullough had left with the truck in the morning, October 7th. The same evening, October 7th, four Ea were hanging still, big, motionless, strongly pulsating in the sky, three to southeast and one to northwest. One Ea appeared later, as so often out of nowhere, as it were, directly over the Observatory.

My Log Book tells me that DOR had made us miserable that same day 11:00 a.m. I removed DOR by Orurization at 5:00 p.m. The sky went blue-gray again. A note in my Log Book reads as follows: "Today, 10/7/54 some people feel as if drawn out of energy, suddenly, and stopping suddenly. The mountain ranges are lusterless, black, the sparkle gone that had enlivened the landscape during the past two weeks or so, when NR-OR was used (ORUR). There is no doubt that I am at war with Ea. What seemed only a possibility one year ago, is certainty now. They either 'hide' among the stars, or they move on same pathways, i.e., they are in same fixed position with respect to earth (as are the fixed stars). The switch-like, sudden sensation of being drawn from or being relieved was reported by many workers."

"*October 8/9/1954*: 4-5 Ea in sky last night again, beginning 10:00 p.m., with 'drawing off' felt. One moved slowly to north, then upward, appearing and disappearing. Southern red ball not there after 9:00 p.m. I operated twice, 20 seconds each at 0.6 Oranur with good result. Two days' dull blackness gave way this morning to stronger, distant blue and sparkle. Ea was ineffective, it appears."

Two Air Force planes searchingly circled the observatory that same day.

The Second OROP Ea

On *October 10, 1954,* about 19:00 hrs. we noticed a large red object in the sky to the west, low above Bald Mountain. We orurized the atmosphere and drew directly from the Ea a few minutes after 19:00 hours, one minute. To our amazement *the Ea moved from the first to the second position in the following sketch* from Log Book, p. 50, that is, to the *south*.

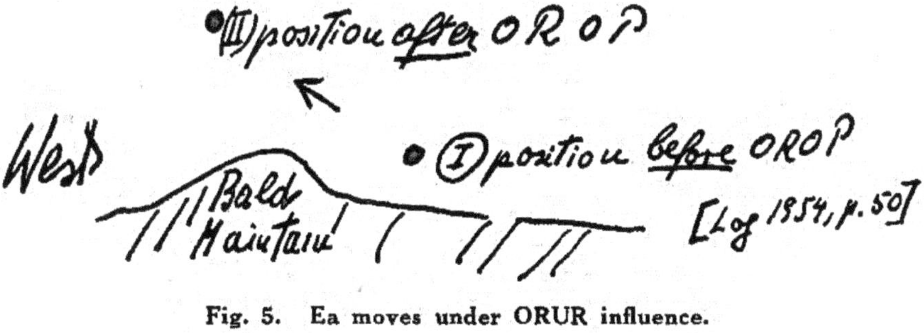

Fig. 5. Ea moves under ORUR influence.

The Ea had also become smaller, less red and it was somewhat higher in the sky. It certainly was neither a planet nor a fixed star according to this behavior. Operator Eva Reich saw it later sink down beneath the horizon.

A second, yellow object to the west at approximately 30° up faded out at 19:30 hrs. after a two minute direct

Orur draw. It came back faintly twice and finally was no longer visible with binoculars. We had the distinct impression of a struggle, fading alternating with strong flashing pulsation, wobbling, moving in various directions. It moved south to north while struggling:

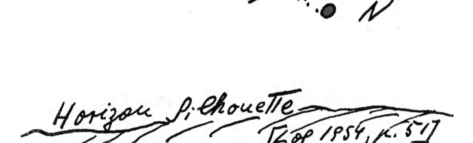

Fig. 6. A second Ea is affected.

It disappeared after weakening, waning and blinking. It returned and went out again 19:53 hrs., came back to luminate once more at 19:55, became fainter, smaller, as if further away in the west. At 20:08 all of them *at the same time*, 1 south, 1 north, 1 west, 1 northwest seemed to remove themselves, growing smaller and fainter, *as if by common command*.

I was about making these notes in my Log Book in my study with my lab watch timing the events when I heard at 20:11 hrs. Eva Reich and William Moise exclaiming on the observation deck: *"A flare!"* A flare had just streaked down eastward from zenith. I had not seen the flare myself that evening but I saw many at other times myself. They were abundant in Tucson the night Orur was flown in, according to reports by those present at the airport (see Tucson report, p. 199) upon arrival of the ORUR material.

At 21:35, one hour and 24 minutes later the Ea to north and south were again there; they had climbed higher in the sky, against the regular star movement. Ea to west and northeast were no longer visible.

This operation Ea was phoned through after conclusion, to our secretary, Mrs. Helen Tropp, in the nearby Rangeley village. A note in my Log Book, p. 52, 1954, reads as follows:

"*Tonight for the first time in the history of man, the war waged for ages by living beings from outer space upon this earth (with respect to DOR, Drought and Desert, WR, 1956) was reciprocated with ORANUR with positive result.*"

The days had been replete with shaking events. Fortunately, I had become used during Oranur to such events and I had trained a few students well enough to stand by me without running away, as so many had done before under the pressure of DOR.

I wish to quote from the First Report on Oranur, 1951:

"We may assume that the OR energy ocean which fills all space is the carrier of the vibrations related to light. However, the relationship seems to be a much closer one. *The OR energy unit itself,* as it develops from and sinks back into the OR ocean, emits light, *strongest and sharpest at the peak* and *weakest during the period of rise and fall.*

"Careful examination of the dots on photograph (Fig. 7a, below) reveals several most interesting details:

"1. Most of the dots are *black,* only a very few are white, i.e., corresponding to effects of ordinary light.

"2. Every single one of the black dots has a sharp 'center' or 'core,' and a less sharp periphery or 'field.'

"3. The intensity and the size of the single dots vary greatly.

"4. Some of the white dots show a sharply defined black center.

"It is advisable not to interpret all these details at once. A major mistake made at this moment might well jeopardize a correct explanation for decades. We can, however, coordinate one definite characteristic with what we already know about OR energy functioning:

"1. The units of OR energy are not rigidly equal. There are not two units exactly the same in size or intensity.

"2. Each dot shows, if well developed, a 'core' and a 'periphery,' the former always more intense than the latter (Fig. 7b, below).

"3. The white field around some of the black dots points to a luminating area around the OR energy unit. This is exactly what we see with our eyes in the darkroom: **The luminating centers have a luminating 'aura' of lesser intensity.**

"This coordination of the physical with the psychological observation forms a sound foundation for further investigations.

"The black and white dots we see on the photograph also agree with the theory we have tentatively built to comprehend the functioning of the cosmic OR energy. *What we see are most likely the peaks of the single units.* If we cut one of the well formed single units crosswise, we can easily see that the **point** is surrounded by a less luminating **field.**

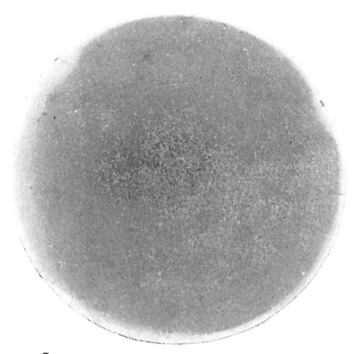

a

Orgone Energy Units
from Earth Bions on photographic plate (1944)

BLACK "CORE" = μ

WHITE "FIELD" = N

b

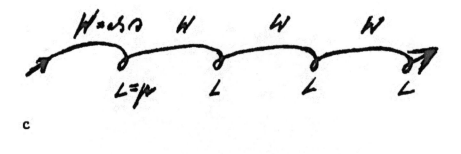

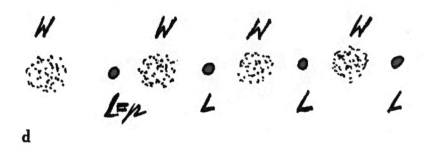

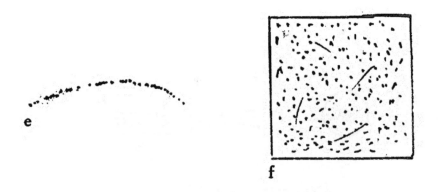

Fig. 7. OR energy photograph and charts

W .... Wave
L .... Loop

"We wish to stop at this point. Further research will probably amplify these first theoretical formulations.

"Let us summarize our results so far:

"1. The primordial substratum, the cosmic OR ocean, is moving in an undulatory fashion and in a certain direction in our planetary system from west to east as a whole, faster than the planetary globe.

"2. Out of this undulating substratum innumerable single concentrations of OR energy emerge, comparable to sharp crests of single waves of greatly varying intensity and extensity (Figs. 7c and d).

"3. Both the OR energy ocean and the single OR energy units luminate.

"4. The natural color of the general substratum is bluish-gray or bluish-green sky, ocean, protoplasm, bions, etc.; that of the concentrated units is deep purple or violet. Gross, streak-like concentrations appear whitish-blue and are rapid, in contradistinction to the other types of OR movement (Figs. 7e and f).

"5. The formation of concentrations to single distinct units follows upon excitation of the OR energy ocean in various ways: Presence of other orgonotic systems, electromagnetic sparks, metallic obstacles, and, foremost, nuclear energy.

"6. The basic character of all these phenomena is of a *functional* nature. There is nothing mechanical, rigid, or absolutely identical in it. Yet, there is clearly a CFP at work, a common law which governs all distinct units, may the variations be ever so manifold. Each single unit possesses a sharp 'core' and a less sharp 'periphery.' This agrees with the bio-energetic structure of every living organism and also with planetary systems. They, too, are composed of a core and an energetically weaker periphery.

"Each single unit passes, accordingly, through four typical phases:

"1. *Birth* through *concentration* of a certain amount of primordial energy.

"2. *Rise in energy level* through further concentration: '*Growth.*'

"3. A sharply luminating **peak**, most closely allied to a point of light.

"4. *Decline* and *death;* the unit merges again with the substratum. Thus, birth and death, growth and decline, the CFP of all living and nonliving nature, seem to be preformed already in the basic functioning of the single, tiny OR energy unit. *Each unit is a unique, unrepeatable event. Yet all orgone energy units follow a common law of functioning. Lawfulness and endless variation are thus not incompatible opposites;* they are paired functions of the CFP of nature in general.—May, 1950"

(**The Oranur Experiment**, First Report, 1947-1951.)

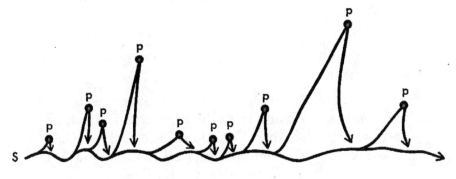

S .... substratum of primordial cosmic energy
p .... points or peaks, luminating

Birth and Death of Orgone Units

Fig. 8

Conclusions Regarding Influencing Ea with ORUR - Spacegun

(1) The energy equilibrium of Ea can be disturbed or even put out of order by withdrawing energy from it directly. Its behavior is directly related to the successful operations.

(2) Signs of Ea having been affected: At first *stronger blinking* as if due to reactive struggle; *fading,* repeatedly interspersed with stronger lumination; getting smaller possibly with *field shrinking;* appearing to *recede* into greater distance; *fading out completely* with or without reappearance of lumination (see photos, pp. 20, 21); summary simultaneous disappearance of many Ea as if on common command; *flares* of warning (?) dropped from sky as if in alarm; *erratic movements* out of order.

(3) The power of the Oranur Spacegun is tremendous due to the sensitivity of the OR energy ocean. Early regulation and coordination of its use is paramount to planetary safety from abuse by Higs.

Due to the long, drawn out orurization on October 10, 1954, it rained gently on the morning of the following day.

The court in Portland was informed that same day that the activities of the Orgone Institute Press in sending out our literature, enjoined by Higs, would be resumed. Our Ea work was infinitely more important than an order which had been inspired and unlawfully obtained by Moscow Higs and was issued by a well-meaning but misled Judge. (See WR: "Atoms for Peace vs. The Hig," *Address to the Jury, History of Orgonomy, A-XII—EP, Supplement 3,* text of letter to Judge Clifford verbatim.)

That same day, too, I reached an understanding of how Orur worked; why the counts soared so high when an otherwise inactive ORUR was put into heavy lead shielding. In the open, without metal shielding, ORUR was peaceful, inactive, corresponding to the foggy type of existence. When put into the lead container, the field was apparently restricted and ORUR changed from the foglike to the *pointed* existence as seen on the lighted green screen and on photographs. This activity was responsible for the high GM action up to 100,000 CPM. It was the *restriction* of its peaceful activity which caused its *"warlike"* behavior. Why had I not seen this simple logic in the behavior of the Life Energy sooner? A peacefully living wild animal, if forcefully restricted, would suddenly act exactly the same way, or rather the OR energy within the animal acts the same way as ORUR within lead.

All biological motion, inner as well as locomotion, appeared now to be the reaction of OR to restriction of its free motility; it changed from the *foglike* to the *pointed* existence. The luminating points (see "Oranur Experiment, First Report," p. 195 and Fig. 8, p. 42 above) are *mechanical discharges* and thus constitute the source of a *motor force*. The action of the OR *motor* belongs here: An even sequence of impulses, registered on the GM counter, at a rate of at least 3,000 per minute, sets a motor into motion (see "Orgone Energy Bulletin," 1948). Also, a leg gone "asleep," wakes up again with sensations of ants crawling or fine needles pushing toward the skin surface. Good theory formation began to pay off. The assumption was fruitful in that it permitted further development to take place in the Arizona desert. It deserves further special, careful study.

The northwestern yellow Ea, the same which had faded out upon direct orurization, did not reappear in the sky after the "battle" of October 10th. There was only a red Ea ball hanging low, 5-10 degrees up, in the western sky on the northern shoulder of Bald Mountain. However, the DOR emergency continued to be critical at Orgonon. My dog, Troll, had eaten little for three days. On the 13th, he was only lying around, sick and miserable. I, too, could not eat because of nausea. My daughter felt dizzy, as if "drawn from the head." She ate little. Grethe Hoff reported by phone from Boston that her little son had been sick for two days and that he was not eating. Our bath water was black-greenish after use that day, a certain sign of much DOR having been soaked from our bodies.

The Ea seemed very active lately. The strongest draw effects were felt from the southern one. I could not rid myself of the idea that, if there were any spacemen around or if these yellow and red luminating balls represented some kind of machines directed energetically from afar, whoever directed them must have been madly minded. That same day my GM counter failed completely by fading and jamming when the Orur joined the lead container. I did not understand why, except that the activity must have been very high, too high for the device to take.

Only a few days were left for us at Orgonon. Moise and my son Peter left Orgonon for Tucson, Arizona, two days after the "battle," on October 12, 8:15 a.m. My own departure was scheduled for October 19th.

Summary of Available Data, October, 1954

This was the state of knowledge available to the expedition:

(1) *Increasing the energy potential in the atmosphere* by way of ORUR was now available in addition to the old technique of *lowering* the potential at will by way of drawing off energy with the Cloudbuster.

(2) *"Decrease"* and *"Increase"* of potential could be interpreted in terms of functional physics simply as a change in the form of the atmospheric energy from the *foglike* (unexcited, low) to the *pointed* (excited, high) state

of existence and vice versa. Only the "pointed" state of OR registers on the GM counter.

(3) There was little doubt left: One could now reach far into space with ORUR: the range was limitless theoretically, since the OR energy ocean is endless and most sensitive to stimuli as demonstrated by the processes of dawn, dusk, and our actual operations over vast stretches of space.

(4) The change from the "cloudy" to the "pointed" state of existence, or *activation* of OR was now possible by simply *impeding its freedom* of *"lazy"* motion or by direct irritation such as friction, sparking secondary coil systems, nuclear material, heat, etc.

(5) At that date, due perhaps to the increased DOR Emergency, we felt that there could be no doubt left as to the purposefulness of the activities of Ea: Energy was being withdrawn from our planet, with the consequences known now, 1956, far and wide as *"DOR-Emergency"*; decay of vegetation, the crumbling of granite rock, a feverish atmosphere. OR energy laws, mostly unknown to us earthmen, were used technically in the Ea operations. We earthmen had only just begun to make contact with these technological matters. We had, in addition to all this, to fight against the malignant activities at our backs of frustrated, dorized, neurotic individuals of our own kind. These were unmistakably represented at that time by the irresponsible activities, the conspiracy of corrupt, ignorant officials, the Higs in the FDA. *"Orgone Energy does not exist,"* they said to the Judge, while the planet was being attacked with the use of this very same Cosmic Energy. It is little consolation that no one will remember these Higs only a few years hence.

(6) All of us who partook in this emergency felt "The war of the Universe was on," with Orgonon against its intention or design playing the role of a "Planetary Valley Forge," as it were.

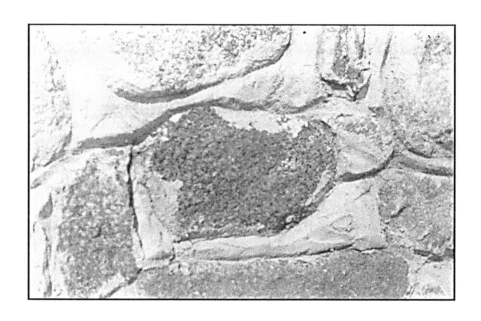

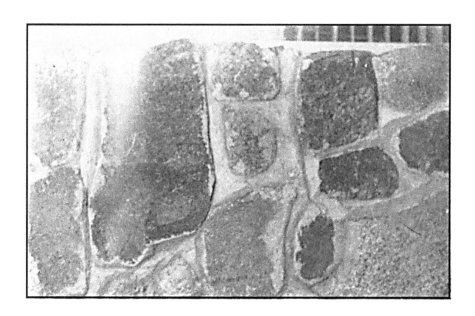

Fig. 9. Melanor on Observatory rocks, two views

CHAPTER III
CONTACT WITH THE U. S. AIR FORCE
Introduction

I shall now introduce certain activities of the U. S. Air Force as far as they were connected with Ea on the one hand and our OROP Desert Ea work on the other hand.

The U. S. Air Force is the natural organization in the western world responsible for the interplanetary developments to come. Although all other military services are involved like every single living being on this globe, the Air Force carries a special responsibility. It operates in the atmosphere and watches the frontier upward toward outer space. It is also the organization which had already some warlike contacts with visitors from outer space. (See the AAF-ATIC report by Ruppelt, 1956.) It maintains a most elaborate system of study of the Ea problem. It is therefore only natural that the Air Force became interested in my work with primordial cosmic energy. This contact was most crucial in the development of the desert work. I shall reveal as much as is necessary for the responsible planetary citizen to know about the planetary emergency. Not all can be revealed, of course. The problem is grave, new, devoid of any historical precedent, loaded with responsibility in addition to grave, unsolved problems of natural science. Only a crank or freedom-peddler will complain if some matters are kept secret for some time to come.

First Contact with the USAF

On January 28th, 1954 I happened accidentally to observe two bright yellow-orange lights moving in front of a mountain range downward toward a lake some two miles north of my "lower house" at Orgonon in Maine. This observation was reported to the Air Force:

ARCHIVES of the ORGONE INSTITUTE

January 29th, 1954

Mr. Lester Farrar, State Trooper
Rangeley, Maine

Dear Mr. Farrar:

We would appreciate it if you would forward the enclosed letter to the Air Force Intelligence in our vicinity. We do not know the address/

Sincerely yours,

ORGONE INSTITUTE
Secretary

ARCHIVES
of the
ORGONE
INSTITUTE

January 29th, 1954

To
U.S. Airforce Intelligence

1192 Ec.

Gentlemen:

    I have been asked to forward to you the following report on observations made by Dr. Wilhelm Reich and myself on the evening and during the night of January 28th to January 29th, 1954.

    At about 22 hours, the following was clearly visible with unaided eyes as well as with binoculars from the North window of the lower house at Orgonon, Rangeley, Maine: A brightly shining light was seen moving in the direction of Badger's Camps to the North, moving in the direction West to East among the trees, and disappearing as it passed behind the trees and appearing again as it reached spaces between the trees. When the light was behind the trees and not directly visible, a bright shine illuminated clearly the surrounding atmosphere. If there was ever any doubt about whether such moving lights were stars, mistaken for moving objects, this time any such doubt was removed in the following way: The bright, white-yellowish light was moving in the valley in front of Spotted Mountain, or, in other words, with Spotted Mountain as its background. The object seemed headed toward Round Pond, and later on this shining glimmer was to be seen further to the right, or to the East respectively. Within a few minutes, a second, similar, lighted object was seen again coming into the same region from the left or the West, moving toward the right or the East among the trees in the valley with Spotted Mountain as background. It was the same appearance and disappearance and the same shining glimmer. Only the second time, the shining object was moving slightly lower than the first one. There was no doubt whatsoever, to judge from these appearances, of the reality of this observation. This time "mistaken stars" could be eliminated with certainty, since the object was not moving as usual among stars, "out of formation" as it were, but it was found in a region where no stars could possibly ever be.

    Miss Ilse Ollendorff reported in the morning of January 29th, 1954, that about 12 Midnight (she did not look at the clock), she observed a similar, but brighter and bigger because closer object to the East of Dodge Pond, hovering in front of Saddleback Mountain

Air Force Intelligence   1/29/54   page 2

somewhere near Dallas Hill or closer. It was seen rising once vertically upward, settling down again and disappearing.

There was no fear or excitement connected with these observations which seem to confirm somewhat subjectively doubtful similar observations reported to you as observed in January 15th, 1954.

                              Sincerely yours,

                              Ilse Ollendorff
                              ORGONE INSTITUTE
                              Secretary

(handwritten addition in original):

These details were dictated by me this morning. Factual observations as well as theoretical considerations are removing any doubt as to the <u>reality</u> of "visitors", no matter whether from outer space or from elsewhere.

There is much more to be said about this in a different connection.
January 29th, 1954
                    signed: Wilhelm Reich
                                        M.D.

Flight 3-G, 4602D AISS
Presque Isle AFB, Maine

F3G   10 March 1954

Dr. Wilhelm Reich
Orgone Institute
Rangeley, Maine

Dear Sir:

In reference to your letter of 29 January 1954 to Major John Dunning, Dow Air Force Base, pertaining to the unidentified sightings witnessed by yourself and Miss Ollendorff, subject officer notified this organization to take whatever action necessary, since this unit is interested in investigating unidentified aerial phenomena which have been observed and reported.

Our headquarters in Colorado Springs, Colorado, was notified immediately and informed of the situation. This unit was then instructed to forward to you the USAF Technical Information Sheet, Form A, (reference inclosure), so that your observations could be fully and accurately recorded. This questionaire is self-explanatory, and no difficulty should be encountered in answering questions and making descriptions.

It is requested that this form be returned to this unit as soon as possible after completion. The questionaire will then be forwarded to higher headquarters for thorough evaluation.

Thanking you for any and all consideration extended, I remain,

                                        Steven J. Hebert
1 Inclosure   STEVEN J. HEBERT
  USAF Tech Info Sheet   1st Lt., USAF
  (Form A0   Asst OIC, Flt 3-G

FLIGHT 3-G, 4602D AISS
PRESQUE ISLE AFB, MAINE

F3G   16 March 1954

Dr. Wilhelm Reich
Orgone Institute
Rangeley, Maine

Dear Sir:

In reference to telephone conversation between Miss Ilse Ollendorff and 1st Lt. Andrew Matyas on 16 March 1954, a copy of the USAF Technical Information Sheet is inclosed for your action.

1 Incl
  USAF Tech Info Sheet

ANDREW MATYAS
1st Lt, USAF
OIC, Flt. 3-G

ORGONE INSTITUTE

March 19th, 1954

Reference: Ea

1st Lt. Andrew Matyas, OIC., Flt. 3-G
4602 D AISS
Air Force Base
Presque Isle, Maine

Dear Lieutenant Matyas:

I am returning the completed questionnaire on "Unidentified Objects" to you for your further disposition.

A manuscript, containing a survey of the orgonomic view on possible visitors from outer space, is awaiting publication. In view of the urgency of the problem, and the responsibilities involved, I am enclosing this manuscript for further clarification of the problem as seen from orgonomic basic research of the past few years.

Under separate cover pertinent publications will be forwarded to you.

Sincerely yours,

Wilhelm Reich

Wilhelm Reich, M.D.

( ) omitted

Form A

U. S. AIR FORCE TECHNICAL INFORMATION SHEET

This questionnaire has been prepared so that you can give the U. S. Air Force as much information as possible concerning the unidentified aerial phenomenon that you have observed. Please try to answer as many questions as you possibly can. The information that you give will be used for research purposes, and will be regarded as confidential material. Your name will not be used in connection with any statements, conclusions, or publications without your permission. We request this personal information so that, if it is deemed necessary, we may contact you for further details.

1. When did you see the object?
   28 Day **Jan.** Month **1954** Year

2. Time of day: **10 - 10:15** Hour Minutes
   (Circle One): A.M. or **P.M**

3. Time zone:
   (Circle One): a. **Eastern**
   b. Central
   c. Mountain
   d. Pacific
   e. Other _____

   (Circle One): a. Daylight Saving
   b. **Standard**

4. Where were you when you saw the object?
   ORGONON Nearest Postal Address **P.O. RANGELEY** City or Town **MAINE** State or Country
   Additional remarks: _I was observing the sky from my study in "Tower House" in darkness, using Binocular 7x50, N° 7147, TOZUNA, in routine manner_

5. Estimate how long you saw the object. ___ Hours **15** Minutes ___ Seconds

   5.1 Circle one of the following to indicate how certain you are of your answer to Question 5.
   a. **Certain** (timed)   c. Not very sure
   b. Fairly certain   d. Just a guess

6. What was the condition of the sky?

   (Circle One): a. Bright daylight
   b. Dull daylight
   c. Bright twilight
   d. Just a trace of daylight
   e. **No trace of daylight**, "Earthshine" +++
   f. Don't remember

7. IF you saw the object during DAYLIGHT, TWILIGHT, or DAWN, where was the SUN located as you looked at the object?

   (Circle One): a. In front of you
   b. In back of you
   c. To your right
   d. To your left
   e. Overhead
   f. Don't remember

55

8. IF you saw the object at NIGHT, TWILIGHT, or DAWN, what did you notice concerning the STARS and MOON?

   8.1 STARS (Circle One):
   a. None
   b. A few *Toward North*
   c. Many
   d. Don't remember

   8.2 MOON (Circle One):
   a. Bright moonlight
   b. Dull moonlight   *¼ moon*
   c. No moonlight — *pitch dark luminescence of atmosphere*
   d. Don't remember

9. Was the object brighter than the background of the sky?

   (Circle One):   **a. Yes**   b. No   c. Don't remember

10. IF it was BRIGHTER THAN the sky background, was the brightness like that of an automobile headlight?:

    (Circle One) **a. A mile or more away** (a distant car)?
    b. Several blocks away?
    c. A block away?
    d. Several yards away?
    e. Other

11. Did the object: (Circle One for each question)

    a. Appear to stand still at any time?   Yes   **No**   Don't Know
    b. Suddenly speed up and rush away at any time? Yes   **No**   Don't Know
    c. Break up into parts or explode?   Yes   **No**   Don't Know
    d. Give off smoke?   Yes   **No**   Don't Know
    e. Change brightness?   **Yes** No   Don't Know
    f. Change shape? *at first some-what*   **Yes** No   Don't Know
    g. Flicker, throb, or pulsate?   Yes   **No**   Don't Know

12. Did the object move behind something at anytime, particularly a cloud?

    (Circle One): **Yes** No Don't Know.   IF you answered YES, then tell what it moved behind: *single trees silhouetted against dark sky background, later*

13. Did the object move in front of something at anytime, particularly a cloud?

    (Circle One): **Yes** No Don't Know.   IF you answered YES, then tell what it moved in front of: *"Spotted Mountain" on dark background and disappeared in valley toward "Round Pond" i.e. Eastward (see sketch)*

14. Did the object appear: (Circle One): a. Solid? b. Transparent? **c. Don't Know**

15. Did you observe the object through any of the following?

    a. Eyeglasses   Yes   No   e. Binoculars   **Yes**   No
    b. Sun glasses   Yes   No   f. Telescope   Yes   No
    c. Windshield   Yes   No   g. Theodolite   Yes   No
    d. Window glass*   **Yes** No   h. Other *in open air*

    * no "reflection" — no lights in neighborhood

16. Tell in a few words the following things about the object.

   a. Sound _none (inside building; heard hum on other occasions)_

   b. Color _bright yellow - white, at times toward orange_

17. Draw a picture that will show the shape of the object or objects. Label and include in your sketch any details of the object that you saw such as wings, protrusions, etc., and especially exhaust trails or vapor trails. Place an arrow beside the drawing to show the direction the object was moving.

   Spotted Mountain as background
   _22:00 hr_ → _visible behind trees_ →
   [See also attached sketch]   _shine only_ →
   _22:15 hr_

18. The edges of the object were:

   (Circle One): a. Fuzzy or blurred   e. Other _surrounding atm sphere_
   b. Like a bright star _but much_   _luminated in valley._
   c. Sharply outlined _larger_
   d. Don't remember

19. IF there was MORE THAN ONE object, then how many were there? _two_
   Draw a picture of how they were arranged, and put an arrow to show the direction that they were traveling.

   Both moved as described under 17, one ~~after~~ at a time. The first had disappeared in valley when the second appeared where the first had ~~been~~ among the trees a few minutes before, at ~~approx. * in~~ 17.

20. Draw a picture that will show the motion that the object or objects made. Place an "A" at the beginning of the path, a "B" at the end of the path, and show any changes in direction during the course.

[See sketch]

21. IF POSSIBLE, try to guess or estimate what the real size of the object was in its longest dimension. _____ feet. not possible to guess, omission preferable

22. How large did the object or objects appear as compared with one of the following objects held in the hand and at about arm's length?

(Circle One):
- a. Head of a pin *(circled)*
- b. Pea *(circled)*
- c. Dime
- d. Nickel
- e. Quarter
- f. Half dollar
- g. Silver dollar
- h. Baseball
- i. Grapefruit
- j. Basketball
- k. Other _____

22.1 (Circle One of the following to indicate how certain you are of your answer to Question 22.
- a. Certain
- b. Fairly certain *(circled)*
- c. Not very sure
- d. Uncertain

23. How did the object or objects disappear from view? Behind the low ridge between Orgonon and Badger's Camp, the ridge being lower than Spotted Mountain north of it.

24. In order that you can give as clear a picture as possible of what you saw, we would like for you to imagine that you could construct the object that you saw. Of what type material would you make it? How large would it be, and what shape would it have? Describe in your own words a common object or objects which when placed up in the sky would give the same appearance as the object which you saw.

No answer possible to this question

About Twice as large as Jupiter when largest

25. Where were you located when you saw the object? (Circle One):
   Owl
   a. **inside a building**
   b. In a car
   c. **Outdoors**
   d. In an airplane
   e. At sea
   f. Other _____

26. Were you (Circle One)
   a. In the business section of a city?
   b. In the residential section of a city?
   c. **In open countryside?**
   d. Flying near an airfield?
   e. Flying over a city?
   f. Flying over open country?
   g. Other _____

27. What were you doing at the time you saw the object, and how did you happen to notice it? *Observing the sky*
   It is part of my routine in Orgone Energy Research to observe (at various times and intervals) the sky, the clouds, the stars, the socalled "DOR-Clouds", sometimes during the night.

28. IF you were MOVING IN AN AUTOMOBILE or other vehicle at the time, then complete the following questions:

   28.1 What direction were you moving? (Circle One)
   a. North   c. East   e. South   g. West
   b. Northeast d. Southeast f. Southwest   h. Northwest

   28.2 How fast were you moving? _____ miles per hour.

   28.3 Did you stop at any time while you were looking at the object?
        (Circle One)   Yes   No

29. What direction were you looking when you first saw the object? (Circle One)
   a. **North**   c. East   e. South   g. West
   b. **Northeast** d. Southeast f. Southwest   h. **Northwest**

30. What direction were you looking when you last saw the object? (Circle One)
   a. **North**   c. East   e. South   g. West
   b. **Northeast** d. Southeast f. Southwest   h. Northwest

31. If you are familiar with bearing terms (angular direction), try to estimate the number of degrees the object was from true North and also the number of degrees it was upward from the horizon (elevation).

   31.1 When it first appeared:
      a. From true North $N - 35°$ ape degrees.
      b. From horizon $+15°$ degrees. $70 - 15$

   31.2 When it disappeared:
      a. From true North $E - 5°$ degrees.
      b. From horizon $- 5°$ degrees.

32. In the following sketch, imagine that you are at the point shown. Place on "A" on the curved line to show how high the object was above the horizon (skyline) when you *first* saw it. Place a "B" on the *same* curved line to show how high the object was above the horizon (skyline) when you *last* saw it.

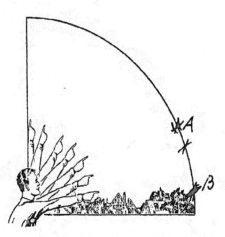

33. In the following larger sketch place an "A" at the position the object was when you *first* saw it, and a "B" at its position when you *last* saw it. Refer to smaller sketch as an example of how to complete the larger sketch.

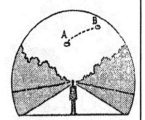

34. What were the weather conditions at the time you saw the object?

34.1 CLOUDS (Circle One)
   a. **Clear sky**
   b. Hazy
   c. Scattered clouds
   d. Thick or heavy clouds
   e. Don't remember

34.2 WIND (Circle One)
   a. No wind
   b. Slight breeze
   c. Strong wind
   d. **Don't remember**

34.3 WEATHER (Circle One)
   a. **Dry**
   b. Fog, mist, or light rain
   c. Moderate or heavy rain
   d. Snow
   e. Don't remember

34.4 TEMPERATURE (Circle One)
   a. **Cold**
   b. Cool
   c. Warm
   d. Hot
   e. Don't remember

35. When did you report to some official that you had seen the object?

   29th / January / 1954   (sent a few days later)
   Day   Month   Year

36. Was anyone else with you at the time you saw the object?

   (Circle One) **Yes**   No

   36.1 IF you answered YES, did they see the object too?
      (Circle One) **Yes**   No

   36.2 Please list their names and addresses: Miss Ilse Ollendorff, Orgonon, Rangeley, Me.
   Miss O. was called in to observe when second object moved in on same route, slightly lower than first.

37. Was this the first time that you had seen an object or objects like this?

   (Circle One) Yes   **No**

   37.1 IF you answered NO, then when, where, and under what circumstances did you see other ones?
   See attached sheets. My attention was called to these phenomena when seen work before I saw similar light objects except behind our observatory, or

38. In your opinion what do you think the object was and what might have caused it?

   See attached sheet

answer to 37.1 : Similar phenomena were called to my attention over the preceeding months on several occasions, especially on January 13th and 14th, around 7:34 PM; also doubtful observations of similar kind on January 17th and January 19th, 1954. Observation of January 28th was distinguished by mountain background.

39. Do you think you can estimate the *speed* of the object?

   (Circle One)   Yes   (No)

   IF you answered YES, then what speed would you estimate? _____ m.p.h.

40. Do you think you can estimate how far away from you the object was?

   (Circle One)   (Yes)   No

   IF you answered YES, then how far away would you say it was? _1'98 Ri off_

41. Please give the following information about yourself:

   NAME _Reich M.D._   _Wilhelm_   _____
         Last Name   First Name   Middle Name

   ADDRESS _Orgonon_   _Rangeley_   _Maine_
            Street   City Zone   State

   TELEPHONE NUMBER _99_

   What is your present job? _Basic Research, Natural Science_
   Orgonomy

   Age _57_   Sex _male_

   Please indicate any special educational training that you have had.

   a. Grade school _4 years_   e. Technical school _18 years Research Labor_
   b. High school _8 years_   (Type) _Orgone Energy Research_
   c. College _5 years_   f. Other special training _Medicine, Psychiatry_
   d. Post graduate _8 years_   _Physics, Bioenergetics_
      Medical Faculty Vienna, Austria, 1922   see attached list of publications

42. Date you completed this questionnaire: _____ _____ _____
                                             Day Month Year

U. S. AIR FORCE TECHNICAL INFORMATION SHEET
(SUMMARY DATA)

In order that your information may be filed and coded as accurately as possible, please use the following space to write out a short description of the event that you observed. You may repeat information that you have already given in the questionnaire, and add any further comments, statements, or sketches that you believe are important. Try to present the details of the observation in the order in which they occurred. Additional pages of the same size paper may be attached if they are needed.

NAME __WILHELM REICH, M.D.__
(Please Print)
SIGNATURE __Wilhelm Reich__
DATE __March 18, 1954__

(Do Not Write in This Space)
CODE:

See attached Survey; further information available upon request

"Ex" is my code word for "Unidentified Objects"
Please confirm receipt of Ex-survey and photostatic copy (Ex)-formulae in writing.
This communication will be published at some future date, if no objection will be raised by responsible AAF official

WR

Enclosures
1 sketch
1 map
1 manuscript copy "Survey"
1 Photostatic copy

Table of Events concerning EK
11067, 11073,
11108, 11114, 11147, 11148, 11150, 11151,
11153, 11156, 11158, 11165, 11165, 11166, 11167
11169, 11170, 11171, 11176,
11182, 11192, 11205, 11230, 11247, 11263
11204,
11270-73, 11225,

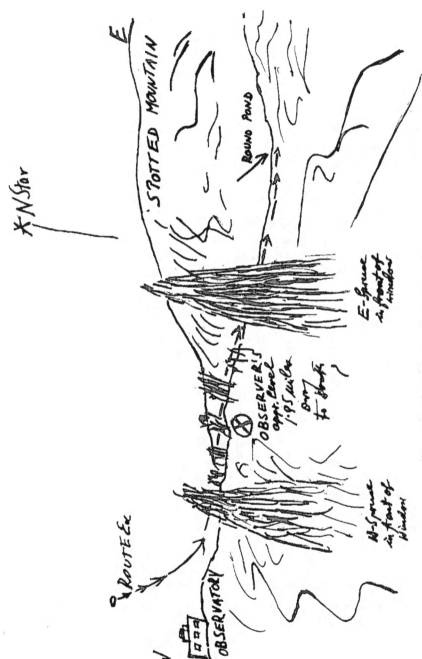

Fig. 10. Ea in front of Spotted Mountain

On March 17th, 1954, two days before the order of Injunction was issued in Portland, Maine, a letter was written to the nearest Air Force base:

Avc. Vel.

March 12th, 1954

Commander
Flight 3-G, 4602D AISS
Presque Isle Air Force Base
Maine

Dear Sir:

On March 12th, 1954, we received a questionnaire regarding unidentified objects in the Rangeley Region. The accompanying letter was signed by 1st Lt. Steven J. Hebert, but the letter was not written on official stationery.

Since the answer to the questionnaire involves major responsibilities, we would appreciate it if the request would be repeated on official Air Force stationery. May we also ask you to send us another set of the questionnaire, since we would like to keep one completed copy for our files.

Sincerely yours,

ORGONE INSTITUTE
Secretary

The same day I forwarded to the AAF a "Survey on Ea" Archive No. AA 11277. The equations contained in this report had been worked out a decade before around 1944. The equations had freed the problem of gravitation from g, the symbol of mass attraction. The gravitational function had been reduced to wave length, λ, and frequency of oscillation, s, or, its reciprocal value, the period per oscillation, t.

Also, a massfree equation for swinging energy had resulted from orgonometric calculations.

These equations were now transmitted to the Air Force Intelligence of the U. S. A. in full cognizance of their impact upon the future of the planet. To "sell" them would have appeared like asking a top surgeon's fee of a man fatally injured in an automobile accident.

Survey on Ea

Here follows the "Survey on Ea," verbatim, quoted from the original manuscript:

AA 11277

App. put 36a-36?

SURVEY ON OR (March 17, 1954)

(Ex .... ORgonomic term for "unidentified objects")

It is impossible to tell in a few pages the whole story of orgonomic research in the realm of cosmic functioning. Instead, I shall tell how I became interested in the "Flying Saucer" problem.

Ten to fourteen years ago, 1940 to 1944, soon after I had discovered the OR energy in the atmosphere, the cosmic implications gradually became evident. Physical pendulum experiments (1944) led - with the true spontaneous logic of objective functions - toward orgonometric equations concerning numerical relations in the cosmic OR energy functions. The result was a pendulum formula ($1t^{-2} \neq 100$) and a mass free energy formula $\neq$

$(E_\sigma \neq \mu \cdot W \neq \Delta \cdot C \cdot \Delta \cdot \rho \neq \Delta^3 \sigma^2)$

A manuscript in German was written in an attempt to work out details. The formulas and the manuscript together with some charts were put away into the archives. The reason was this: I did not believe mathematical formulas to be worth publishing unless they were confirmed by <u>physical observations and experiments</u>. I had acquired this attitude when I discovered the physical phenomena of atmospheric and cosmic OR energy which fills the universe; then I compared my factual observations with the purely abstract, mathematical formulations which had emptied space, declared the "ether" to be non-existent and - as a makeshift on a narrow factual basis - not only had blocked further factual physical research, but also stood in the way of the true comprehension of the cosmic primordial energy.

The crucial importance of the subject for the understanding of "<u>Visitors from Outer Space</u>" may excuse my blunt language.

Thus, I waited many years for further factual developments. These developments began to show up when, after a pause of three years (1944-1947), I

SURVEY

started my work on vacua as described in the Oranur experiment ("Vacor"). The OR energy motor, too, was then in the process of being worked out. It was published in the first number of the OEB, in 1949.

I knew of the existence of a "saucer" problem around 1950. But I never had paid much attention to it. The idea of visitors from outer space did not at all seem strange to me. Why should only the Earth be inhabited by intelligent beings? There was no reason whatever to assume such a thing, except because of ones vanity as a homo divinus or sapiens; solipsism on the part of man was too well known from the phantastic cruelties of an age, only 4-500 years ago, which thought the world to be flat and also to be the center of the whole universe, with man as its focal point of divine attention.

The fact of space ships coming down to Earth was therefore nothing "new". I had heard about it on the radio at times. But it was never in the center of my attention. This, inspite of the fact that friends had told me of saucers having been seen over Orgonon in 1951. Even, when in August of 1952, standing on the front porch of my summer house at Orgonon, I heard something buzz by from horizon to horizon, SW to NE, within a few seconds, I paid little attention to it.

My interest in the saucer problem was seriously awakened in the following manner:

We had suffered gravely from a black substance which had poured down over Orgonon since April 1952, the so-called "Melanor". It was obvious from all apearances that this Melanor (see "The Blackening Rocks", OEB,1953) came from somewhere in the Universe. But only in November 1953, when I first read the Keyhoe Report, did things begin to fall into place.

Disclosures regarding the Flying Saucers, such as noiselessness, bluish shimmering lights, rotating discs underlying their motion, fell into place

SURVEY 3.

with some of the facts I knew well from cosmic OR energy functioning; functions such as silent operation which I had experienced when the OR motor had been worked on in 1947-1948. Or blue lumination in 0.5 micron pressure vacuum tubes. The old orgonometric functions came back, one by one, especially those which had dealt with the gravitational ~~formulas~~ *equations*.
<u>The facts confirming these abstractions seemed to be at hand now</u>, in a strangely <u>practical</u> manner. Things tied in further. Function after function coordinated itself with what I had read about the "unidentified objects". The final picture as of today is about this:

1. The "<u>CORE MEN</u>" (CORE = COSMIC ORGONE ENGINEERING), as I came to call them, apparently were thoroughly conversant with the laws of functioning in the cosmic OR energy ocean, especially *with* gravity as a function of superimposition.

2. They use cosmic OR energy in propelling their machines.

3. Their "blue lights" were in agreement with the blue color characteristic of all visible OR functions, sky, protoplasm, aurora, sunspots, depth of moon valleys as seen at dusk, the color of OR energy lumination in vacor tubes, etc.

4. The changes of color from blue to white or red, etc. , I knew well from various studies of OR phenomena and I had seen some practically in vacor tubes (see Oranur Experiment, page 249 f).

5. The CORE MEN were obviously riding their space ships on the main OR energy streams of the Universe (see COSMIC SUPERIMPOSITION on "Galactic and Equatorial Stream").

6. The tremendous speeds which they were able to achieve were not in disagreement at all with the tremendous quantities I had calculated for the OR energy streams of the Universe in 1940 to 1944. (There are still many gaps

there, and many uncertainties. The tables of the kr^x number system are in my possession. See also attached copy).

7. A speed of 10 or 15 thousand miles per hour did not appear impossible in the light of these numbers; on the contrary: It appeared quite natural.

8. They rotated their discs in harmony with the OR waves they rode on. Rotating discs describe exactly what I had calculated 10 years previously as the so-called KEW ("Kreiselwellen")or SPINNING WAVES, without any knowledge of space ships actually riding cosmic OR waves. From these waves, I had derived my ~~mass free energy formula~~ *equation* $E_\sigma \neq cb^3 n^2$

Things were fitting well, even too neatly for my taste. Therefore, I hesitated to tell anything to anyone about them. I only worried about what might have happened to the facts and ~~formulas~~ *equations* which a student had acquired in 1947 to 1949. He had worked on the OR energy motor, and he disappeared in 1949 under mysterious circumstances. He was under suspicion of having been a Russian spy. His case was urgently transmitted to the FBI in 1949, and again in 1950. We urged again in 1953 to ascertain his possible role in the case. ⚹

9. Just as space is not empty, light does not "come down to us from the stars and the sun". It is an effect of lumination in the OR energy envelope of the planets. <u>It is a local phenomenon.</u> Therefore, there is theoretically no limit to speed in cosmic space, only technically. This agrees with the apparently limitless quantities in energy functions which characterize the orgonometric kr^x system as progressing in geometric proportions (kr^x System).

The accompanying summary of basic equations was sent to the National Academy of Sciences sometime in 1948. Since industrial interests in pharma-

⚹ *Note, April 1956: Lately, serious considerations forced the idea upon me that W.W., the man who suddenly disappeared in 1949, may have been somehow connected with optical space ships. He was intercepted by the A.E.C. See Protocole of Telephon conversation N° 105, "Conspiracy", 8.16.1948.*

ceutic chemistry were rampant in attempts to destroy Orgonomy ever since 1937. I have no way of knowing what happened to this information.

The office of the President of the USA has been kept informed on all major developments ever since 1951, when the Oranur Experiment went under way.

March 1954

WILHELM REICH, M.D.
Orgone Institute
Orgonon
Rangeley, Maine

(The handwritten portion above reads as follows: "Note, April 1956: Lately, serious considerations forced the idea upon me that W.W.,* the man who suddenly disappeared in 1949, may have been somehow connected with actual spaceships. He was intercepted by the AEC. See protocol of telephone conversation, No. 105, 'Conspiracy,' 8/16/1949.")

---

* William Washington

BASIC ORGONOMETRIC FUNCTIONAL EQUATIONS

(deposited and legalized at the
Law Office Arthur Garfield Hays,
New York on January 13, 1948.)

(1) **General Functional Equations:**

  a) $x \not{f} y$

  b)
  $$A \not{f} \begin{cases} x \not{f} \begin{cases} a \leftrightarrow \\ b \leftrightarrow \end{cases} \\ y \not{f} \begin{cases} c \leftrightarrow \\ d \leftrightarrow \end{cases} \end{cases} \text{etc.}$$

(2) **Basic Functional Transmutation:**

  $l \not{f} m$

(3) **Primordial Cosmic Orgone Energy Function:**

  a)
  $$E_\sigma \not{f} \begin{cases} 1^3 t^{-2} \not{f} \begin{cases} l_1{}^3 t_1{}^{-2} \\ l_2{}^3 t_2{}^{-2} \end{cases} \\ 1^3 s^2 \not{f} \begin{cases} l_1{}^3 s_1{}^2 \\ l_2{}^3 s_2{}^2 \end{cases} \end{cases} \to \text{etc.}$$

  b) $E_\sigma \not{f} \begin{cases} mc^2 \sim \text{(Waves)} \\ h.v... \to \text{(Quanta)} \end{cases}$

  c) $1^3 t^{-2} \not{f} \begin{cases} D^3 T^{-2} \\ D_1{}^3 T_1{}^{-2} \end{cases}$

  Wilhelm Reich

  d) $l_{kr}{}^3 x \cdot e_{kr}{}^2 x \not{f} K \cdot kr^2 x$ (see leg. Document of March 22., 1944)

73

e) $1 \not= 10^2 t^2$

(4) kr^x - Number-System:

$$kr^x \times 10^{2x} \begin{cases} rkr \times 10^{xa} \\ kr^x \times 10^{xb} \end{cases}$$

(5) "Gravitation":

a) $g_{kr} \not= \pi \not= r^2\omega \not= 1t^{-2}$

b) $1t^{-2} = 100$
$1t^{-2} = 10^3 \rightarrow 1^3 t^{-2}$
$10^2 \pi^2$
$r^2\omega \rightarrow D^3 T^{-2}_Q \rightarrow E_Q$

(6) One Special Application:

$$2rkr \times 10^7 \text{eS} \begin{cases} 365 \cdot 2425 \text{ED} \begin{cases} 2rkr \times 10^0 M_Q \\ 2rkr \times 10^3 E_{Q\sim} \text{eS} \end{cases} \\ 2rkr \times 10^4 E_Q \begin{cases} 2rkr \times 10^4 Z^e \\ 2rkr \times 10^5 R'' \end{cases} \end{cases}$$

Newyork
Jan. 10, 1948

Wilhelm Keal

Received Jan 13, 1948
Julian G. Culver

Two Orgonometric Propositions

① General:

$$N \to \begin{cases} R'' \to \begin{cases} R_a \\ \to nSr \end{cases} \\ 2° \to \begin{cases} \to ntr \\ T_a \end{cases} \end{cases}$$

② $N_{Earth} \ni H \ni 2h \ni k_r{}^4 \times 10^4 \text{apec.}$
$\quad\quad\quad\quad\quad\quad\quad\quad\quad\quad\quad\quad\quad\quad \ni GY_\sigma \to$
$N_{Sun} \ni P_{krx} \ni Y \ni H_{krx} \ni k_r{}^4 \times 10^1 QT$

New York, Jan. 11, 1948

Wilhelm Reich

Received Jan 13, 1948
Julian H.␣␣␣␣␣

Specification of N

$kr^x = 4^x$; $x = 1, 2, 3, 4$ etc., KRW.. $kr^{3x} \times 10^{x}$
$S = 390625$; $t = 256 \times 10^x$
$A = 92160$
$B = 9350208$
$h = 65536$
$h+ = 6574365$ ⎫ (f Planck's Number)

$H_{kr^x} = 1314873$
        5259492
        84151872 ⎫ Galactic Numbers: Earth

$X \times H = 33660748$

$P_{kr^x} = 2359296$
        9437184 ⎫ "   "   Sun

$X = 256$.

These Numbers are part of the transmutation of classical to ozonometric astrophysical calculations. Since Rotation and movement in space of the Sun are simultaneous, the Copernican "circles" and the Keplerian "Ellipses" become invalid, while the numerical values remain. The classical functions are

Received Jan 13, 1941
  Julian H. Lenhers

transformed orgonometrically into periodic functions such as of pendulums which swing and move in space simultaneously: "Cosmic Pendulum". The classical numbers corroborate this unavoidable transmutation numerically. The functional Equations serve this purpose. They are not fully calculated yet, but the basic work has been done beginning 1939-1940 when the atmospheric Orgone Energy was discovered.

Wilhelm Reich

New York,
Jan 11, 1948

The foregoing mathematical calculations were received by me in the office of Hays, St. John, Abramson + Schulman, 120 Broadway, New York 5, N.Y.

Julian G. Cuchman

Sworn to before me this 13th day of January 1948
Tierney De Forest

TIERNEY DE FOREST
Notary Public, State of New York
Residing in Kings County
King Co. Clk's No.571 Reg. No.463-D-9
N.Y. Co. Clk's No.117 Reg. No.705-D-9
Qns. Co. Clk's No.225 Reg. No.175-D-9
Bronx Co. Clk's No.21 Reg. No.257-D-9
Certificate Filed in Westchester Co.
Commission Expires March 30, 1949

(The handwritten portion above reads as follows: "These numbers are part of the transmutation of classical to orgonometric astrophysical calculations. Since rotation and movement in space of the sun are *simultaneous,* the Copernican 'circles' and the Keplerian 'Ellipses' become invalid, while the numerical values remain. The classical functions are transformed orgonometrically into periodic functions such as of pendulums, which swing and move in space simultaneously: 'Cosmic Pendulum.' The classical numbers corroborate this unavoidable transmutation numerically. The functional equations serve this purpose. They are not fully calculated yet, but the basic work has been done beginning 1939-1940, when the atmospheric Orgone Energy was discovered.")

First Direct Contact with the Air Force Technical Intelligence Center (ATIC) in Dayton, Ohio

One of our operators, William Moise, had the order to go through Dayton, Ohio, to make contact there with the chief officer of the Air Force Technical Intelligence Center (ATIC) and to report our operations of October 10, 1954, which resulted in the fading out of "stars" in the sky. He was also advised to give all the information necessary to the understanding of the event.

He phoned on October 11, 5:00 p.m., for an appointment on Thursday, October 14th, late afternoon. He reported to me on October 14, 8-9 p.m., from Dayton, Ohio, that General Watson was not available and had cancelled the appointment. I gave him the directive on October 15, 8:00 a.m. by phone to Dayton to use his own judgment in handling the garbled affair. He told me that a Major Hill had left a message for him in his hotel. Later the same day, Moise reported by wire that a "good discussion" had taken place with the Deputy Commander of ATIC on the Ea event, on Orur, on the problem of transportation of Orur to Arizona. The latter problem was delegated to

Washington. On October 15, he phoned through the basic points of the meeting. The chief officer had excused himself, though he had granted the appointment previously. He was represented by the Deputy Commander, Colonel Wertenbaker, of the U. S. Air Force. The Air Force from then onward was coded in our records as Eaa to match the Ea designating the unknown flying objects (UFO).

The report on that meeting follows:

Report to WR

of conference at Air Force Technical Intelligence Center (ATIC), Wright-Patterson Air Force Base (AFB), Dayton, Ohio, concerning Ea, Oranur Radium (ORUR), and Equations.

10/11/54.............Communication of the Orgone Institute with ATIC.

16:30 (about) Rangeley, Maine, time:

Long distance telephone conversation between *reporting operator* and *General Watson*, Commander, ATIC, Wright-Patterson AFB, Dayton, Ohio.

Wilhelm Reich, M. D., Director of the Orgone Institute, wishes to report to you personally, the following occurrence: *Two UFO's, which the Orgone Institute refers to as Ea, were disabled last night, October 10, 1954, by Wilhelm Reich.*

This is such a crucial and decisive matter that it must be reported wholly and in detail to you personally. Wilhelm Reich requests an appointment with you to report this orally and in detail. We are traveling to the southwest on an experimental expedition into the desert there and, as a representative of Wilhelm Reich, request an appointment with you for this coming Friday, October 15th.

General Watson replied that he could not arrange to see me upon Friday since a group from CIA in Washington was coming Friday. However, he could arrange to see me on Thursday. When I replied that this would be satisfactory, but that since I would be driving I may not arrive till late in the afternoon, General Watson answered that this made no difference, that *if necessary we could continue the conference after supper,* until late in the evening.

General Watson asked me only one question, and that was how did Wilhelm Reich know that the UFO's were disabled? I did not answer this satisfactorily. *I failed to tell him that they went out.* I said only that the manner of their behavior and movement clearly indicated they were disabled.

I further added that Wilhelm Reich also wished to discuss two other questions at the meeting; one concerning a problem of transportation of ORUR, and one concerning the depositing of scientific material with the Air Force.

The arrangements were that I would have an appointment with General Watson for Thursday afternoon, October 14th. I would telegraph on Wednesday to General Watson's office, giving my time of arrival in Dayton. Upon arriving in Dayton, I would telephone his office to find out the time of the appointment.

10/13/54 —— Telegram sent to General Watson, ATIC, Wright-Patterson AFB, Dayton, Ohio, by William Moise from Chambersburg, Pa.:

"Arriving Dayton tomorrow Oct. 14th. Request appointment for 4 pm. Will contact your office on arrival for confirmation."

Orgone Institute, by William Moise

10/14/54 —— Telephone call from Dayton to General Watson's office by William Moise for Orgone Institute, 14:30 hrs.:

4 pm. appointment with General Watson confirmed.

4:00 p.m.-Conference

I was met at ATIC Headquarters by Dr. W. H. Byers, physicist with the Air Force ATIC. Dr. Byers informed me that the previous communications from the Orgone Institute, correspondence, publications, etc., had been directed to him by ATIC, that he was familiar with the work being done by the Orgone Institute.

Dr. Byers then escorted me to a conference room and introduced me to Captain D. M. Hill, USAF, and *Mr. Harry Haberer*, civilian, working with the Air Force in regard to the *history of UFO's*. Dr. Byers then said that General Watson had asked him to extend the General's apologies to me for being unable to meet with me, that unexpected important business prevented him from being at the conference * * *. This news took me by surprise, stunned me. I said nothing. I felt enraged, so said nothing and simply studied the men present. It was clear that a report as crucial and as important as this report could not be presented in the presence of these gentlemen alone. (Dr. Byers is a man with a flabby handshake and eyes that don't look at you.) The Captain looked even younger than I and looked very frightened. Mr. Haberer appeared somewhat bored.

So I told Dr. Byers that my instructions were to present this report from the Orgone Institute to General Watson, that this had been requested by Wilhelm Reich, *that General Watson had confirmed this request.* We had traveled a thousand miles from Maine to Dayton, because we had been assured by General Watson of an appointment. I also said that in my conversation on Monday with General Watson I had been told by him that if necessary the meeting with him could continue after supper late into the evening. I asked them when I could meet with General Watson. Dr. Byers said he didn't know.

I told them that the crucial nature and importance of the report necessitated that it be presented only in the presence of the highest authority. I left my phone number, said that if necessary I could stay on 2 more days to meet with the General and left.

10/15/54:

Captain Hill telephoned me and informed me that he had spoken with General Watson. General Watson once again apologized, said he would be unable to see me, but asked if the Orgone Institute would present the report to Colonel Wertenbaker, Deputy Commander? I said yes and an appointment was made for 09:30.

I was met at Headquarters by Colonel Wertenbaker. I had immediate contact with him. The contact continued and increased throughout the conference. The Colonel told me that General Watson had arrived and taken over command only two weeks ago and that until that time he, Colonel Wertenbaker, had been acting commander.

Present at the conference were Colonel Wertenbaker, Dr. Byers, Capt. Hill and Mr. Haberer.

I first asked Colonel Wertenbaker *if anything had come up since the telephone conversation with General Watson?* I was puzzled since in the conversation, General Watson said that he would meet with me even if it meant doing so after supper late in the evening. The Colonel replied that nothing had come up. Since the Colonel's time was short I requested the other gentlemen to save their questions until after the report, but asked the Colonel to interrupt whenever he had a question. All the gentlemen but the Colonel took notes.

My report is as follows:

The region of the northeastern U. S. A. and Orgonon has been much bothered the last few weeks by Ea's

(UFO's). Some nights there would be 3 or 4 hanging in the sky. They would make the atmosphere black. Dr. Reich, working with the Cloudbuster, would mobilize the atmospheric energy and make it blue again. This struggle had been going on for several weeks. At this point, the Colonel interrupted me and asked me to describe this blackness. I replied that a bleakness would come into the atmosphere, an absence of life, of sparkle, everything looked and felt flat and dead. I added that we could feel the energy being drawn out of us. The landscape looked colorless.

I continued: A few weeks ago Dr. Reich dug up the radium which had been used in the Oranur experiment and which had been buried since that time, a period of about three years. Dr. Reich wished to test this radium, and since has been using it in experiments with the atmosphere, with rain-making, cloud dissipation, successfully.

While doing this, the idea came to Dr. Reich to use it in connection with the Cloudbuster and train it on a UFO. He hesitated in doing this since there was a possibility of them being American craft, but the situation became so bad, the blackness, that he finally decided to.

The following I read from Dr. Reich's *protocol of 10/10/54:*

> 07:00 — A big red Ea on western horizon, very low. Trained Cloudbuster excited with Oranur radium on object for 1 minute. Whereupon the object first sank down, lost its field, faded out and returned as in distress. It changed position, higher and to the south.

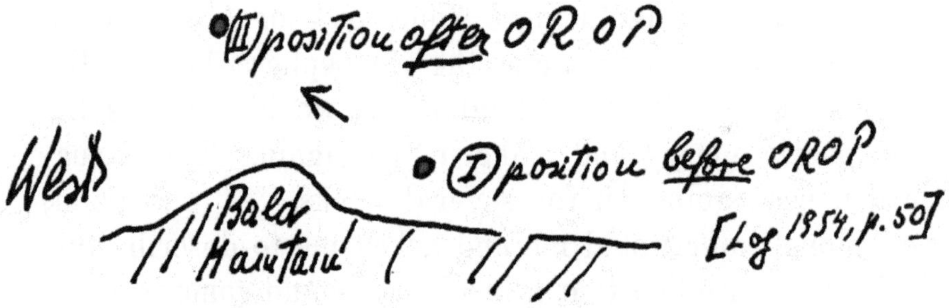

Fig. 11. Ea moves under ORUR influence.

Then later it disappeared altogether.

19:30 — One yellow Ea, a big one, 30 degrees elevation in the northwest.
Oranur trained on this object for 2 minutes, 27 seconds.

19:53 — Object went out for 30 seconds, after pulsating, shrunk. Remained out for 2 minutes.

19:55 — Object came back once more at 19:55, appeared far away, fainter and smaller.

20:08 — All of them, one to the north, one to the south, and one to the west, as if by *"common command"* seemed to remove themselves, fainter and smaller.

20:11 — A bluish-white flare streaked across sky, zenith to the northwest, lasted for 2 seconds.

21:35 — Ea to north and south still there, having climbed into sky, *against star movement*. But those in the west and northwest no longer visible.

22:10 — Dr. Reich *phoned through* the happenings to a secretary in the town of Rangeley.

At the point in this report where the path of the object over Bald Mountain was shown, the Colonel's excitement became high; he asked to see the sketch and studied it for a few moments.

I then reported the May 12th drawing. That on May 12, 1954, between 19:40 to 20:40 after training of Cloudbuster in a certain way towards an Ea in the western sky, the object faded out 4 times.

I continued, speaking directly to the Colonel, and said the Orgone Institute is engaged *in a serious "Battle of the Universe"; that these are critical and crucial matters.* The *Orgone Institute does not like to be put off,* that the *issues can no longer be evaded by responsible authorities,* that the *Orgone Institute cannot take the sole responsibility; it must be shared.* I said, that as they well knew, this discovery and the Orgone Institute are *under attack by little horse thieves, commercial hoodlums, psychopaths, all who want to kill the discovery.* Dr. Reich's skill and experience on the social scene, alone, has prevented this from happening. But in the event that something should happen to Dr. Reich, he has certain scientific material which he wants the Air Force to secure. This material consists of certain space equations. I showed the Colonel the equations which I had brought with me and continued: Dr. Reich feels that although it is doubtful if much can be done with these equations without him, nevertheless he feels the U. S. Air Force is the only agency in the world today capable of the responsibility of possessing these equations. I asked if such a copy as I had with me could be deposited, sealed, with the Air Force to take care of such a situation. *Colonel Wertenbaker* answered immediately that *this could be done and that the Air Force would respect the confidence entrusted to it.* He was extremely serious at this point.

I then brought up the problem of the transportation of the Oranur radium (ORUR). I explained that Wilhelm

Reich was on the way to the southwest for experimental research in the desert region there, *that this was a continuation, a very important part of this Battle of the Universe.* The Oranur radium (ORUR) is an extremely vital tool needed in this battle and needed in the southwest. The problem, which to date we have been unable to solve, is how to transport this radium. The *usual commercial means of transportation cannot be used since* there is no protection against this Oranur radium; on the contrary, it is safest when naked and reacts when shielded or near metal. The only way that it can be transported is to be towed 50 or 100 feet behind a plane. And the only organization equipped to do the transporting is the U. S. Air Force. So the Orgone Institute requests assistance from the Air Force in transporting this material to the southwestern United States. Colonel Wertenbaker said that he did not have the authority to make such a decision. The decision would have to be made in Washington and that he would forward the request to Washington. He asked how soon we wished it transported. I said as soon as possible but I requested that The Orgone Institute get a decision immediately whether or not the Air Force could assist and that the answer should be sent to Dr. Reich at Rangeley before the coming Tuesday since Dr. Reich was planning to start to the southwest on that day. I said that the details of how and when could be worked out later. In discussing the reaction of the Oranur radium when shielded I reported the fact that an SU 5 survey meter gave counts of 80,000 and 100,000 CPM.

Colonel had to leave at this point. He thanked the Orgone Institute on behalf of the Air Force. The contact with Colonel Wertenbaker was excellent throughout the conference. He was serious, intent and looked at me while I talked. He was the only one who did. His excitement increased as the report progressed. I felt he was suppressing his excitement and it is *my impression* that he did not wish to reveal this excitement in the *presence of the Doctor of Physics.*

After the Colonel left, I stayed on with the other gentlemen to answer any questions. There *were few questions.* Capt. Hill asked me to explain further the Oranur experiment. I simply told him that ATIC had a copy of it and that reading it would be better. Dr. Byers asked me if these objects were sighted at places other than the Rangeley region. I told him that they were seen at Hancock, Maine, but that we had no way of knowing about the rest of the country.

The conference was then adjourned at about 10:20.

I received a telephone call at about 12:15 in my hotel. A woman's voice said that she was calling from Wright Field (this is a different field than Wright-Patterson) and asked if I knew how they could reach Donald Keyhoe, that there was a telegram from California for him; they knew he was in this area but did not know where. *The name did not click with me* and I asked her to repeat it and then spell it. At that point, she said, "the former Major Keyhoe who wrote the book on UFOs. We knew you were interested in UFOs, too, and thought you might know where he is * * *." I told her I didn't know, never met him, there was no connection with him * * * sorry I couldn't help her.

/s/ William Moise
Operator, OROP Desert Ea
Arizona, 1954/1955

Technical Research Coordination

Our request to the AAF for help in the transportation of Orur was later denied from Washington, AAF Headquarters. I did not understand why. However, we managed ourselves to get Orur without mishap 3,000 miles across the continent.

I did not know at that time, to quote from the Ruppelt report on UFOs (page 285) that my work was, already in 1952, considered "hot because it wasn't official and the reason it wasn't official was because it was so hot." How much hotter it had become with the Ea event on October 10, 1954 * * *

We received no direct help from the Air Force, financial or otherwise, all through our period of Ea research, trial, suffering and error.

The factual background of this meeting at ATIC on October 15, 1954, is long and complicated:

It comprised many common interests in atmospheric and cosmic research. On July 6th, 1953, the first major successful drought-breaking, rainmaking operation took place at Ellsworth, Maine (CORE, 1954). We knew at that time that our operation had been carefully followed by the AAF. A plane flew close by during the operation. It rained neatly in Oranur fashion on the following night all through from the coast to Rangeley, after a dry spell of about 6 weeks. Two days later on July 8th in the morning, AAF planes appeared over Orgonon trailing what appeared to be research equipment behind them; it was most likely monitoring the background count of the atmosphere. From then onward the planes were around wherever Oranur research was being done, at Orgonon since July, 1953, on the way to Tucson in October, 1954, in the research region around Tucson base all through the winter, 1954/1955, then again over Orgonon during the summer in 1955 and also at Washington during the winter, 1955/1956. This interest of the AAF in our work had its further good reasons.

A second link was given in parallel atmospheric research concerning the behavior of vapor trails emerging from jet engines in a DOR atmosphere. Already in 1952 we had seen AAF planes, jets mostly, crisscross the sky over Orgonon in various directions, always leaving widely visible jet vapor trails behind.

As time passed by, with careful observations of the sky, we began to understand the significance of the behavior of the jet vapor trails. We had never made personal contact with the AAF regarding this matter. But the implications were obvious.

In the first Oranur Report, 1951, I had advanced a hypothesis on the behavior of the atmospheric energy. Years of observation of various states of fog in the mountain areas of Rangeley had shown a close dependence of fog and cloud formation on various states of OR energy: *Clouds would not hold over Orgonon when DOR clouds were hovering in the region.* The clouds would dissipate, split up and drift *around* Orgonon; they would gather again on the other side after having passed by the Observatory. As reported in 1954 in connection with drought, no thunderstorms passed over Orgonon in 1952 and 1953 when the DOR effects were strongest. This meant: *DOR absorbed the moisture from the clouds;* the clouds, after having become fuzzy and thin, dissipated and fell apart. On the other hand, where there was no DOR in strong concentration, where the sky was bright and blue, clouds formed and held easily. They grew by merger or apposition of newly formed clouds.

Now, the same behavior could be seen with regard to the *jet vapor trails. The vapor emanating from the jets would hold together for a long time and over miles of sky when the sky was bright and blue, when DOR was absent. On the other hand, the vapor trails would be thin or would not be visible at all; they would not last long, would dissipate quickly when DOR was heavy.* In very heavy DOR regions the vapor trails would not form at all; they would reappear exactly where blue bright sky would surround heavy DOR clouds.

In this manner a most important, almost exact tool of observation of DOR in the atmosphere had been found by the Air Force. From then onward one was able at will, at any altitude, in any region of the globe, at any time of the day to determine whether DOR was present in the sky or not; one could even judge the intensity of DOR clouds by the degree in which the vapor trails would fail to form or hold, by their thinness, length, etc. The Air Force, so I had observed, had furthermore developed various patterns of jet vapor trails on one and the same level or on different levels of the atmosphere, according to the special research task.

For example, one wanted to know whether DOR was heavier and more concentrated at higher rather than at lower altitude, say more at 30,000 than at 20,000 feet. All one had to do was to send up two jets, one to 30,000, the other to 20,000 feet, to let them fly vertically parallel to each other, and the nature of the trails would tell by direct observation where the DOR concentrations were heavier. This was at least the way I interpreted for my own use with great success the various forms and patterns of the jet trails; it was very helpful in further basic research or in the clarification of pressing Ea problems. Whether the Air Force had actually such problems in mind, I cannot tell.

The DOR clouds, on the other hand, were direct indications of the presence of Ea in the sky, seen or unseen. *DOR clouds and high GM counts have become the two most reliable research tools in the Ea realm.*

I have never spoken to any Air Force scientist about these matters. According to my Log Book protocols, there could be little doubt about the identity of the observations. The Air Force may have developed a different theory; it may have had a different purpose in mind in letting the jets trail their vapors: Whatever the case may have been, these Eaa operations were crucial and became more important as the years passed by for my own Ea research. I assume the AAF was not adhering to this technique over the years without good reason either.

During the Tucson operations, when heavy DOR prevailed, the Air Force would at times let two jets draw a cross of vapor trails into the sky. At the point of crossing, so we learned by way of spacegun operations, an Ea may have been hidden far up, pouring its DOR fumes into our atmosphere. We would then draw from that spot and orurize the atmosphere toward that region of the sky; the atmospheric situation improved within a few minutes for human, animal and vegetation alike. These were great moments in an otherwise rather hectic, harassed existence.

To put forth another example: When the DOR removal operations in the Tucson region were finally successful, prairie grass began growing again; a line of green demarcation, as it were, would spread from our base outward, especially toward the northeast and northwest, beyond which DOR was still hanging heavy, black-gray, low above the landscape, while bluer skies were on the inside of the demarcation line. AAF jets then would follow my car or the truck with the cloudbuster and would test the region for DOR, writing their telling trails into the sky. At the borderline, their trails would more or less suddenly thin out if they flew outward far enough toward the untouched DOR region. It was always a great joy as well as gratification to see those jet trails reach from horizon to horizon when good DOR removal work had been done.

The silent, impersonal cooperation of the Air Force had a good reason. On the Air Force rested the grave responsibility to meet the Ea sooner or later, as the case may turn out, in a friendly or in a war-like manner. Their jets were easily out-distanced by the Ea; the pilots could not keep contact with Ea at gunner's range; they failed to understand how Ea operated. Mechanistic science had no answer. Here, the basically new, fruitful approach of ORANUR became evident. It provided a new viewpoint, rich in facts regarding a motor-force unknown heretofore on earth. It also provided a new mathematical, i.e., orgonometric approach in its *functional* equations (see OEB, 1949-1952, orgonometric equations).

The Air Force knew that a new kind of war was possibly impending. It also learned the hard way, going through many heartbreaking experiences, that present-day mechanistic science had no approach to the Ea problem. Opinions of top administrators, military and civil alike, were heard, to the effect that the interplanetary war may already have begun (a view the correctness of which I personally am convinced); the pressure increased. Thus, it was no wonder that OR research had gained serious attention in high military circles according to the ATIC report. It did not matter then that no one had as yet stuck his neck out in signing a positive opinion about Oranur. Higs * were in 1951 'till 1955 quite busy to destroy Orgonomy. They were equally busy not to let the UFO problem reach a rational solution. UFOs and OR energy *"did not exist"* for mechano-chemistic science.

I had already in 1948 and again in January and March 1954 sent information about the orgonometric aspects of the Ea problem to the Academy of Sciences in Washington and to the Air Force Technical Intelligence in Dayton (see pp. 49-78). The reader will understand why, as long as there

---

* Hoodlums in Government.

are Higs in hiding, ready to steal and to destroy our freedom and existence, I must refrain from telling all I know about the subject; why I must keep some matters secret until the time will be ripe and the situation safe enough, to tell the full story. I do this with regret. It is not *my* fault. I hate secrecy. It goes against my true nature. But I cannot help adhering to it under the circumstances. The report on the orgonomic approach to the technology of Ea was sent to the ATIC in Dayton, Ohio, on March 17th through the Presque Isle Air Force base in Maine. It formed the actual basis of the discussion our operator had with the ATIC on October 15, 1954, regarding our first active contacts with the Ea.

The military interest in Ea was acute; however, it was secondary in importance to the natural scientific interest. The latter was basically mathematical, i.e. orgonometric. However, mechanistic science had, apart from basic principles of learning, little to offer regarding Ea mathematics. Old principles had to be newly applied, new forms of thought had to be constructed *in order to come to grips with Ea,* as new scaffolds have to be put up for a new type of building.

It is only logical that my first serious contact with the AAF should have been a mathematical one. Later on, various branches of knowledge had to be integrated to keep the contact alive in a fruitful manner: observation; psychological judgment on the scene of actual battle; social psychiatry in handling the well-organized, well-hidden, pernicious ways of the Emotional Plague; physics, biology, in addition to geology; farming and atmospheric medicine. Too much it truly was. But our forefathers had to accomplish a similar feat in integration of knowledge on their level at their time in order to survive lesser social crises.

I shall not bother the reader at this point with too much orgonometry. But some facts must be brought forth at this point. It is necessary to show one single crucial difference between mechanistic and functional thinking; a point which will illuminate drastically why Ea turned out to be outside the realm of mechanistic physics as well as mathematics:

On September 4th, 1954, a secretary who rested during lunch on the porch of the lower house at Orgonon, exclaimed suddenly: "A flying saucer, look, look!" I turned my binoculars toward the sky. There were five jets flying in formation toward southeast; they were trailed two or three miles behind by a wobbling, skipping silvery disc which kept pace with the jet formation. Then, suddenly, the disc made a sharp turn and disappeared from sight. It was the first time that I had seen a silvery disc in daytime. It was the wobbling, however, which attracted my attention most. I called Presque Isle Air Force Base and told the officer on duty that their jets had just been trailed by what appeared to have been a spaceship.

I had had several direct and indirect contacts with that Air Force Base previously. A few months before, on March 17th, 1954, I had sent in a form filled out with an extraordinary observation I had made between 22:00 and 22:15 hrs. on January 28, 1954, in the presence of another person (see p. 55). Two luminating objects were seen from a northern window at the lower house at Orgonon moving slowly from west to east toward a northern mountain range, the "Spotted Mountain." This observation was reported to the AAF—on January 29th. The reason for reporting this observation, while other, similar ones were not reported, was a new fact: On January 28th, the two objects were *travelling in front of the mountain*, i. e., *the lights were seen coming slowly down from the sky to the left and passing* to the right into a valley *with the mountains as background behind them*. Such lights had been observed before, but always only in the free sky above the observatory. The most striking feature of the observation was the steady, slow movement, with the mountain as background. The movement was akin to hovering with a slow forward and downward motion.

My report to the ATIC was sent registered (P. O. Rangeley, on March 19, 1954, No. 404). It carries No. 11277 in my Log Book "Table of Events." I had no inkling of the fact that that very same day Judge Clifford in Portland, Maine, had signed and issued the order of injunction which was unlawfully obtained by the Higs of the atomic oil, food and drug business in the U. S. A.: *"Orgone Energy does not exist."* The injunction arrived a few days later, on March 22. My report to ATIC contained, apart from the filled-out questionnaire on the UFO observation, a special paper with information on the orgonometric functions which, so I assumed, underlay the movements of the Ea, titled **"Survey on Ea."** My Log Book tells me that on March 19th, 1954, the functional identity of Ea motion and $\mathcal{E}\sigma$, the cosmic primordial energy motion, was clear to me. An entry 11278 on March 20th registers a discovery on *"negative gravity."*

I shall try to avoid at this point the complicated inner mathematical meaning of this. However, it is necessary to give a survey on the problem.

The "Swing," ⌒

The "Swing" can be easily visualized as the line described in space by a point on the rim of a wheel rotating forward. In relation to the ground, this point on the rim though rotating with even speed in itself, describes a movement of alternating *acceleration* and *deceleration*. In other words, its motion *expands* and *contracts* alternatingly. On the forward turn the point moves faster. On the backward turn it moves slower. The ratio of speed change depends, of course, on the basic speed of rotation: The faster the rotation, the shorter the contraction with respect to the forward motion.

A spinning top shows the same basic function of speed contraction and expansion. The top will move in a more or less curved line at high speed. The line of motion forward will be more even the greater the speed.

At a lower speed of rotation, the pin on which the top rotates will clearly describe a *spinning wave*, a KRW (Kreiselwelle) and *swings* with alternating acceleration and deceleration thus; the number of arrows indicate the acceleration and deceleration:

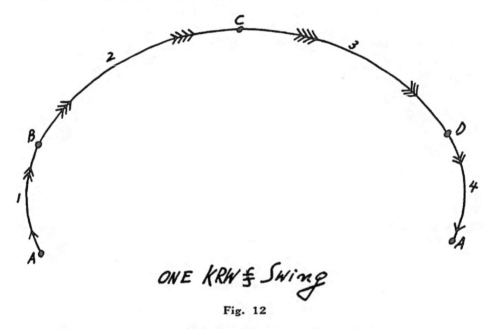

ONE KRW ⹋ Swing

Fig. 12

At the lowest possible speed of rotation the body of the top will wobble, while the Spin will describe short, tight, loops somewhat like this:

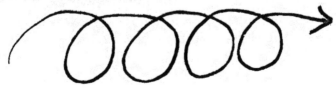

Alternating expansion and contraction of forward motion may also be easily observed in the movement of swinging pendulums under the condition that the point of suspension moves onward in space, while the pendulum body swings:

Pendulum motion → contraction → expansion

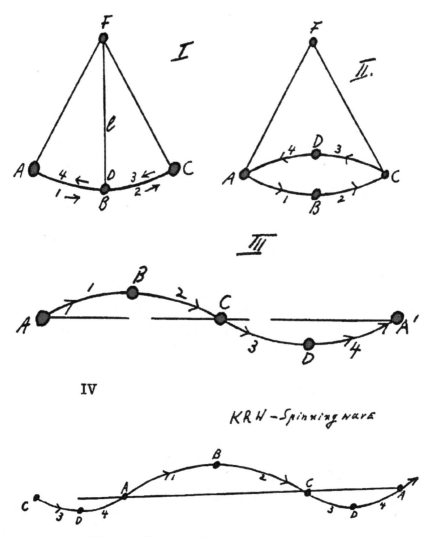

Fig. 13. The forward-moving swinging pendulum

Ea seems to move in straight lines at high speed; it seems to wobble at lower speeds, like a stone skipping on waves. Whether it wobbles like a top when hovering has not been reported in the literature, as far as I know. But its movement is basically **Swinging**.

It may suffice at present to discuss briefly two different approaches to the swing: *mathematically* in terms of known mathematical thought, and *functionally* in terms of orgonometry.

Mechanism freezes the motion in a coordinate system of reference; it analyzes the frozen motion with respect to the chosen framework, the x and y axes. Functionalism, on the other hand, attempts to describe the motion as it is, in order to grasp its lawfulness *without freezing* it. It assumes that there is no special force to move the object. Motion, in terms of OR physics is a basic quality of the OR energy ocean itself. There is nothing that "makes" the cosmic energy ocean move. It moves *"a priori,"* as it were. It makes matter move, just as water waves cause a ball to roll forward on the crests and troughs. The rolling balls always move slower than the wave itself, which "pushes" the ball onward. By the way, the ball rolling slowly forward in the direction of the waves which carry it on, also describes an integrated path of contracting and expanding speed; the ball neither swings straight up and down, nor does it move in a linear fashion forward. Linear movement forward (speed $c = 1 \cdot t^{-1} \neq 1 \cdot s$) is functionally integrated with up and down swings, s. The waves which carry the ball forward function according to the classical massfree wave equation $W = \lambda \cdot s$. The water itself does not move forward, except where it rolls. The waves do move on; that is, energy, not mass, moves on.

Matter does not appear in any of these classical equations. Only wave length (λ) and frequency (s) of oscillations or its reciprocal, the period of time, $t^{-1} = s$, if periodic time, $T = st = 1$, determine speed and wave form: $W = \lambda \cdot s$. *Matter, m, is not contained in the equation.* Now, this is exactly the case in the formulation of the **Swing**, ⌒. The orgonometric equation for the swing, $\mathcal{E}\sigma$, reads:

$$\mathcal{E}\sigma \neq \lambda^3 s^2$$

The Energy of the Swing, ⌒, shows the following characteristics:

1. The *Wave function, W, is functionally identical with the Pulse function, p*. It cannot be otherwise, as the illustration demonstrates:

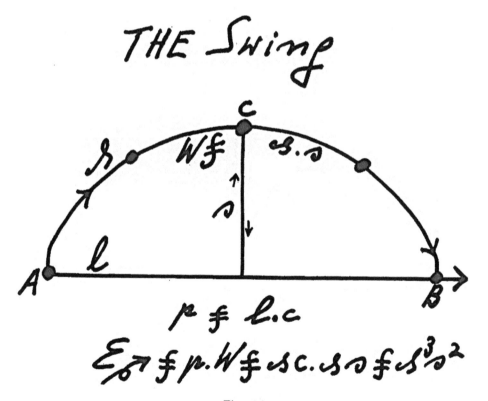

Fig. 14

W ≠ p, or λ · s ≠ 1 · c. Therefore the total energy effect, since 1 ≠ λ, and t^{-1} ≠ s per T = 1 is:

$$\mathcal{E}\sigma \neq p \cdot W \neq \lambda c \cdot \lambda s \neq \lambda^3 s^2 \neq l^3 t^{-2} \neq l^3 s^2 \neq \lambda^3 t^{-2}$$

2. The Energy of the Swing shows a form

   a. Typical of the harmonic law of Kepler,
      $D^3 T^{-2} = d^3 t^{-2}$

   b. Akin to the Planck equation, $\varepsilon = h \cdot v$:

$$\mathcal{E}\sigma \neq mlt^{-2} \cdot 1 \neq mlt^{-1} \cdot lt^{-1} \neq p \cdot c \neq p \cdot W$$
$$(\text{since } c = lt^{-1} \neq W = \lambda s)$$

The total functional coordination yields the following apparently lawful orgonometric form for $\mathcal{E}\sigma$ with respect to Kepler's macrocosmic and Planck's quantum law:

$$E \nearrow \begin{cases} l^3 t^{-2} \nearrow \begin{cases} D^3 T^{-2} = d^3 t^{-2} \\ h.v = \varepsilon \end{cases} \\ l^3 s^2 \nearrow \begin{cases} \lambda.[l.t^{-1} = c] \\ \lambda.s \end{cases} \end{cases}$$

3. The Swing, with T = 16, the total periodic time being divided in 16 parts, and the angular velocity per T/16 measured yields an interesting graph.

   It reminds one with its shape of the Electrocardiogram, the EKG, with its P, Q, R, S, and T points. Thus, once more, the Life Energy, that rules the heart function, demonstrates its origin in the Cosmic Energy Ocean:

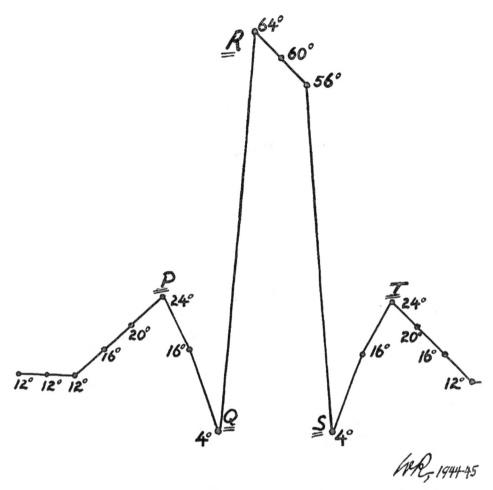

Fig. 15. Basic Form: "Electrocardiogram" is result of change in angular velocity of the Swing.

$$l \neq 100\ t^2$$

The following manuscript (1945) on the *Orgonomic Pendulum Law* was transmitted to the CIA in April, 1956, shortly before the trial in Portland, May 5-7, 1956, following my unlawful arrest in Washington on May 1, 1956 (see "Battle for Truthful Procedure," in Reply Brief for WR vs. U. S. A.). This pendulum law opens the way to gravitational equations without g.

ORGONOMETRY   OROP DESERT Ea

Three Steps:
① ϕhr = +
② ϕhr = 0
③ ϕhr = ÷ (ϕ+)

III. The Orgonotic Pendulum Law [1945]
by Wilhelm Reich

The orgonotic pendulum law proceeds from the following work hypothesis: that the atomic weights *(φ+φ-functions)* of hydrogen, helium and oxygen (1, 4, and 16) can be <u>functionally</u> replaced by numerically identical pendulum lengths of 1, 4, and 16 centimeters. A detailed grounding of this work hypothesis will be presented in another connection.

Atomic weights express gravitational <u>mass</u>, thus, at any given time definite attractive forces of the function of gravitation.

$$G = mg$$

On the other hand, oscillating pendulums are *functional* configurations of <u>oscillatory energy</u>, which are independent of mass m and whose frequency depends exclusively upon the length of the pendulum;

$$g = lt^{-2} \quad (cm/sec^2)$$

in which l does not represent the length of fall in space, but ~~represents~~ *(at a definite point)* the pendulum length; g is constant = k. Orgone physics was forced to adopt the concept that mass m at one time developed out of the <u>mass-free</u> energy of the primordial/*cosmic* orgone. Now since the product of pendulum length and frequency of oscillations per time unit is constant, since further the frequency of oscillations depends on the pendulum length, we express the mass function of gravitation in terms of the oscillatory function, when we apply the set of weight measures 1, 4 and 16 as <u>measures of length</u> in the pendulum experiment.

This preliminary communication must be content with the indication of a broad and complicated relationship.

Let us name the number 4.....kr. The geometric progression of the number $4 = kr$ is accordingly kr^x. x is the series 1, 2, 3, 4, 5, ...x. "kr^x" denotes the kr^x- <u>number system</u>, i.e. the geometric progression of the number 4.

Empirical measurements are *easily* available for the oscillations of pendulums with the lengths 4, 16, 64, 256, and 1024 cm. The number of oscillations for the rest of the kr^x - pendulums can easily be determined.

As a control for this pendulum experiment we use the arithmetic series (*denotes*) of the number 4, thus (n x kr) in which n equals ~~equals~~ 1,2,3,4,5....etc. The arithmetic series of the number 4 we name the __kr - number system__.

The carrying-out of the measurements on Long Island, New York at sea level and around 45° latitude gives the following values for the kr^x - pendulums:

kr^x - Lengths ℓ	(Schwingungen) Oscillations per minute = 64 Sec. / s	Square of Oscillations s^2	$L_{kr^x} \times S_{kr^x}^2 = K_{kr}$
$kr/4$	1	320	102400	102400
kr^1	4	160	25600	102400 ⎫
kr^2	16	80	6400	102400 ⎬ Experi-
kr^3	64	40	1600	102400 ⎬ men-
kr^4	256	20	400	102400 ⎭ tal
kr^5	1024	10	100	102400

On the other hand the values for the pendulum of the kr-system are the following:

14

kr - length	Oscillations per 64 sec.	Square of oscillations	l x s²
1 (kr⁰)	320	102400	102400
4 (kr¹)	160	25600	102400
8	110	12100	96800
12	90	8100	97200
16 (kr²)	80	6400	102400
20	70	4900	98000
24	64.5	4160.25	98846
28	59	3481	97468
32	56	3136	100352
36	52.5	2756.25	99225
40	50	2500	100000
44	47	2209	97196
48	45.5	2070.25	99372
52	44	1936	100672
56	42	1764	98784
60	41	1681	100860
64 (kr³)	40	1600	102400
256 (kr⁴)	20	400	102400
1024 (kr⁵)	10	100	102400 (kr)

The product (lxs²) lies inexactly but still clearly in the vicinity of the value for g ≈ 980 cm/sec² of classical physics multiplied by 10².

The pendulum lengths 25 and 100 cm give experimentally:

15.

l	s	s^2	$l \times s^2 = k$
25	64	4096	102400
100	32	1024	102400

The equation $l \times s^2 = k$ holds exactly only for the kr^x - pendulums and for the ($kr^x \times 10^{2x}$)-pendulums. All other pendulums, the pendulums of the (kr x n) series ~~have a constant that~~ fluctuate are within definite limits and ~~is~~ thus inexact with regard to k.

We ~~characterize~~ *designate* from now on the kr^x - pendulums as "LAWFUL" pendulum lengths, in distinction to the "unlawful" pendulum lengths of the n_{kr} - system. The lawful pendulums have as a *exact* ~~precise~~ constant the product of the pendulum length and the square of the frequency. Further, they possess a numerical harmony in the structure of the kr^x - system. The unlawful pendulums have inharmonious numerical relationships; they do not function in the framework of the kr^x - system.

The function of the kr^x - pendulums can be described as:

$$l_{kr^x} \times \overline{s}_{kr^x}^2 = K_{kr^s}$$

in which Kkr^s = 102400 is valid for *these* oscillations per minute = 64 seconds, and Kkr = 100 is valid for *single* oscillations per second, and Kkr = 25 is valid for double swings per second.

The "orgonotic seconds pendulum" is a certain pendulum length ~~of~~ *in* the kr^x - system. *It* amounts to 100 cm. This pendulum strikes seconds with each movement ~~at about~~ *ca.* 45° north latitude and at sea level.

The pendulum experiments in my laboratory in Forest Hills, New York had established the fact that the constant $l \times s^2/k$ is ~~precisely~~ *exactly* valid only for the "lawful" org-pendulums. With this established, I undertook the following experiments at Jones Beach on Oct. 18, 1944 between ten and eleven a.m. with 69% relative humidity and 75° F. temperature:

1.   A 100 cm. long pendulum with a pendulum body of one

gram weight was set in motion and the number of oscillations measured per 64 seconds. The result was the following: 32 double swings = 64 single swings had a time of oscillation (T = s x t) of 64 seconds in three consecutive experiments. The time measurements were exact within $\pm$ 0.1 - 0.2 seconds.

2.   A pendulum, 64 cm. long with a pendulum body of 4 grams weight gave:

   a. in 64 seconds

      by measurement 1.: 40 double = 80 single swings exact up to 0.0 sec.

      by measurement 2.: precisely the same result.

   b. in 60 seconds

      by measurement 1: 38 double- or 76 single swings in 61 seconds;

      by measurement 2: 37 double swings in 60 seconds.

Only the results 1 and 2a are in harmony with the kr^x - system. Result 2b clashed with it.

The constant, related to a 1 minute = 64 seconds, is equal to $1 \times s^2$ = 102400. Related to 1 second the constant is equal to $1 \times s^2$ = 100 for single swings and 25 for double swings*for all measurable kr^x - pendulum lengths.

These results are in contradiction with the estimates of many text-books of physics which estimate the length of the seconds pendulum as ca. 99.4 cm. This contradiction also exists in the relationship of the pendulum length, l, to the gravitational acceleration, g, for the following reasons. Let us place the numerical values for the classical and the orgonotic seconds pendulum opposite one another:

---

* The distinction between single and double swings (between a half and a whole sinus wave) is significant in another connection.

Classical seconds pendulum		Org-seconds pendulum
Length, l:	99.4 cm	100 cm
Time of oscillation, t, for double swings; for single swings:	2 sec. 1 sec.	2 sec. 1 sec.
Frequency s per 60 sec.:	30 double-s 60 single-s	Frequency per 64 sec. 32 double s 64 ~~double~~ single s
Measuring time $T = s \times t$:	1 min. = 60 sec.	1 min. = 64 sec.
Gravitational acceleration:	$g = 4\pi^2 l T^{-2}$ $= 4 \times 9.8696044 \times 99.4/4$ $= 981... cm/sec^2$	$g_{kr} = 4\pi^2 l T^{-2}$ $= 4 \times 9.8696044 \times 100/4$ $= 986.96044 cm/sec^2$

The acceleration of fall g_{kr} of the kr^x system at sea level and ca. 45° north latitude equals

(1) $g_{kr} = 986.96044 cm/sec^2 = 100\pi^2$

The relationship of the pendulum length l to the gravitational acceleration g remains the same: $l/g = 99.4/981... = 4\pi^2/g_{kr}x = 100/986.96.. = 0.1013...$

Since also t and π remain the same, the equation

$$g_{kr} = 100\pi^2$$

is exlusively an expression of the lengthening of the minute time-measure T^m from 60 to 64 seconds. The measure of time $T = 64 = 4^3$ sec. is a strict kr^x - system-number.

T+0 ∮ FUNCTIONAL EQUATIONS

The theoretic significance of the "functional" equation $g \mathcal{f} 100\pi^2$ lies in the fact that <u>free fall is expressed in terms of a circular function</u>. This fact is significant in astrophysics and requires a detailed presentation which will be given in another paper.

From the pendulum equations

$$g_{kr} = 4\pi^2 . l . T^{-2}$$
$$g_{kr} = l . T^{-2} = l s^2 = l/100$$ for the org-seconds-pendulum, and
$$g_{kr} = 100\pi^2$$

follows both mathematically and functionally:

$$4\pi^2 \cdot \ell \cdot t^{-2} = 100\pi^2$$, therefore is
$$\ell \, t^{-2} \neq 25 \text{ cm/sec}^2 \text{ (for double swings)}$$

where ℓ this time = <u>pendulum length</u> (not length of space as in free fall) and t ~~characterizes~~ <u>period</u> of ~~duration~~ oscillation of a double swing. We obtain the functional equation

(2) $\quad \ell \neq 100 \, t^2 \quad$ (for single swing)
$\quad\quad \ell \neq 25 \, t^2 \quad$ (for double swings)

experimentally from the measurable kr^x pendulums and we dispose of the old, classical equation $l \, t^{-2} =$ ca. 981 (or in the org-system = ca. 987) cm/sec^2.

In the case of the ~~oscillating~~ (κx) org-pendulum ℓ characterizes the <u>length of the pendulum</u>. In the case of the equation for free fall (ℓ) ℓ characterizes <u>gravitational acceleration</u> per second.

In the equation of gravitation $g = 1 \cdot t^{-2}$ dyn $\sqrt{\text{cm/sec}^2}$ =987, a <u>SPACE</u> - function is expressed.

In the pendulum equation $1 \, t^{-2} = 100$ cm/sec^2, on the other hand, a <u>TIME</u> - function is expressed, since oscillating pendulums represent cosmic clocks.

Thus we obtain a <u>functional relationship between two functions</u>, in which length is expressed ~~multiplied~~ periodic by time squared:

(3) $\quad 1 \, t^{-2} =$ ca. 1000 cm/sec$^2 \quad\quad \ell \, t^{-2} = 100$ cm/sec^2
$\quad\quad\quad$ SPACE - FUNCTION $\quad\quad\quad\quad$ TIME - FUNCTION

The functional equation $1 \neq 100 \, t^2$ can be applied in cosmic functions. ~~As has already been proven,~~ I reserve the presentation of this fact for a later time. It cannot be given within the confines of this short preliminary communication for it requires elucidation of the essence of ~~FUNCTIONAL~~ EQUATIONS. $(x \neq y)$
Orgonometric

19.

We obtain a <u>second</u> important functional relationship, if we functionally equate the ~~force~~ function of WEIGHT
$G = m \times g = m \times l \times t^{-2}$ in dyns with a second ~~force function~~ the moment of rotation $K = mr^2\omega$. That we need to postulate such a functional identity will also be shown in another more detailed exposition:

$$G \downarrow f\ K_{\Theta}\ \text{or}$$
$$mg\ f\ mr^2\omega$$
(4) therefore $\ g \downarrow f\ r^2 \omega\ \Theta$

In this case the gravitational acceleration g is represented by a <u>circular function</u> or <u>rotatory function</u> ($\downarrow f \Theta$)

- The functional equations presented above were obtained about four years ago in the course of the determination of the orgonometric functions of the <u>rotating orgone envelope</u>. They express a close ~~narrow~~ functional relationship of the gravitational acceleration to the cosmic orgone, especially curve to the rotation of the orgone envelope of the planet earth. They have already demonstrated their astronomical usefullness in a wider compass.

The theoretical significance of these ~~functional~~ orgonometric equations lies in the fact that they place us in the advantageous position of being able to investigate the function of gravitation and the gravitational acceleration <u>independent</u> of g. The gravitational acceleration g is, stridtly speaking, a <u>local</u> function which is visible and accessible to us only on the surface of the earth. Since the direct observation of natural phenomena was from the very start a rigorous demand of ~~orgonomy and~~ orgonometry, we can now enjoy the great advantage of <u>submitting functional equivalents for the function g in which g no longer appears.</u>

We are still faced with the task of ~~strictly proving~~ whether and to what extent the functional equations

(2) ~~$Kr^2\ l_0 f$~~ $100 t^2$ and
(4) $\downarrow l\ t^{-2}\ f\ r^2 \omega\ \Theta$

are valid in natural research.

Orgonon, August 12, 1947

According to several inquiries it has been ascertained that the handwriting on the preceding pages in the mathematical equations is Mr. Myron Sharaf of Boston (Brookline) who has helped translating and the William Washington in typing the manuscript on the Pendulum experiment.

Febr. 20, 1956 Willh. Reich

(The handwritten portion above reads as follows: "According to several inquiries it has been ascertained that the handwriting on the preceding pages in the mathematical equations is Mr. Myron Sharaf's of Boston (Brookline) who had helped William Washington translating and in typing the manuscript on the pendulum Experiment.")

CHAPTER IV

DOR CLOUDS OVER THE U. S. A.

The Trip from Rangeley, Maine, to Tucson, Arizona, October 18 to October 29, 1954 — 3216 Miles

(*First reported in CORE, vol. VII, Nos. 1-2, March, 1955*)

I departed from the clear bio-energetic atmosphere at Orgonon with some misgivings, wondering what reactions to the energy atmosphere elsewhere to expect in the organisms of workers who had been exposed to the various phases of Oranur for years. Orgonon had become a kind of oasis. Only the day before departure, I had noted the unusual springlike quality of the atmosphere at the end of October: The buds were swelling on the birch branches, and strawberry blossoms were blooming at the steps of the Orgone Energy Observatory.

In the Cloudbuster, and other material, we carried tools to modify new environments, to clean out DOR and bring in fresh energy from the infinite Orgone ocean wherever we went. But one could not be 100 percent sure that functions, which applied on the East Coast, also apply to the Western Deserts.

I gathered valuable data *driving through* rather than *flying over* the continent, confirming many aspects of the theory of desert development which had resulted over the past three years from the work done in the laboratory concerning Oranur (*pre-atomic*) Chemistry, Melanor and Orite.

The observation of DOR in the atmosphere, and its effects on living things, vegetation and population, took up full attention for the entire distance of the trip. Within the general DOR layer covering the land there appeared zones worse or better, and a few isolated areas actually were DOR-free at the time of our passage, but generally DOR was found everywhere.

Over towns and cities it hung as a blackish, low smokey-looking pall. Where large cities were closely spaced, one could drive for hours without meeting zones of clear energy in between. Arriving from a relatively fresh region, I summarized this fact as the result of "Large cities + chemical offal + decaying nature + Ea + DORized health officials * * *." It was a *fact* that was easily available by DOR-removal engineering operations applied to the "smog" problem of big cities (see report on OROP INFANT, this issue). Neglect by public servants of such basic means for tackling public health problems of millions in a constant DOR emergency was no longer possible. One had to differentiate between their blocking of new insights by *doubt* (*rational*) and by *neurotic* mechanisms (*irrational*).

It was about 196 miles from Rangeley, Maine, that we first definitely entered a sharply delineated zone of DOR (a zone extending about 50 miles north of Boston, Mass.). From Boston onward, except for a few exceptions, the landscape represented well-known signs of clearcut DOR DESERTS: Brown-black, disintegrating, crumbling rocks, trees dried, branches bent to the ground like rubber hoses, foliage lacking autumn coloration and turned brown, leaves crumbling to brown powder in one's hand, CPM (on the Geiger Counter) of an erratic nature (anywhere from 50 to 200), square, fuzzy, gray-blackish drought clouds overhead.

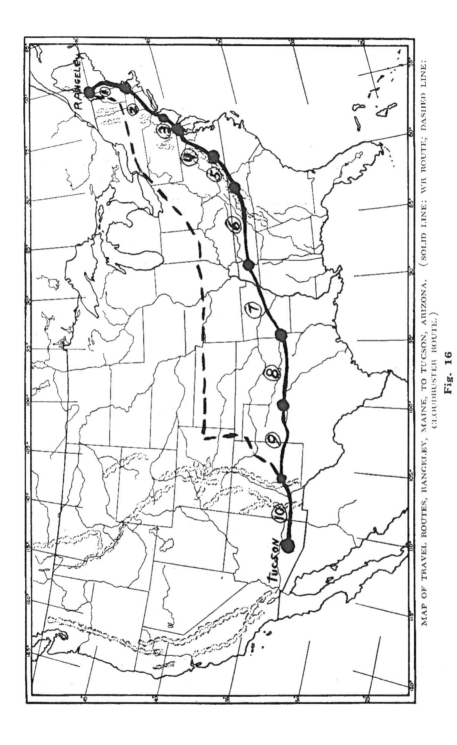

MAP OF TRAVEL ROUTES, RANGELEY, MAINE, TO TUCSON, ARIZONA. (SOLID LINE: WR ROUTE; DASHED LINE: CLOUDBUSTER ROUTE.)

Fig. 16

Some examples from the notes I dictated will show how similar the general impressions were:

"Hartford, Conn. (9 AM, 10/20/1954): DOR++++; Air gray; Trees very bad, decaying; Bitter taste; Clouds in irregular bands * * * also mottled."

"Milford, Conn.; Rocks DOR-attacked * * *."

"Near Baltimore, Maryland (10/21/1954): Trees clearly drought stricken, bent, broken, disintegrated. No autumn coloration, leaves turn brown * * *."

"Warrentown, Va. (10/23/1954): Blackness, sudden sharp DOR zone, region desolate, earth parched, branches without leaves, withered, bent, droopy—total impression: *dying countryside* * * *."

The destruction of trees was general throughout the East. In New England trees were snapped off along their lower trunks. The cross sections of the splintered trunks appeared bone dry, brittle, no sap was present. My notebook comment was: "The Weather Bureau attributes the DOR destruction of the Maine woods to a hurricane which wasn't there (Edna)." For the cracking of the stems was basically due to their weakened condition, the result of the drying out of the living trees by DOR, which had preceded Hurricane Edna by two years.

About 50 miles from the heart of New York City (10/20/1954 at 9:30 AM) near Westport, Conn., for the first time a distinct change was noted: The atmosphere against distant hills now appeared bluegray, a wind arose, the trees moved, the clouds rounded and thickened, lost their squareness, and no DOR was felt. I concluded and later was able to confirm this impression that "someone was drawing." A short time before, Dr. Michael Silvert had conducted a brief DOR-removal operation from a site at the East River, New York City. This operation freshened the atmosphere for a few hours only, since it became hot and dry again when we crossed into New Jersey around noon.

It was in New Jersey that we first encountered the devastating effects which the prolonged drought had caused. The ground had become parched; trees drought-stricken. Frequent were "ghost trees" whose brittle branches had broken off near the trunk, leaving a few naked stumps reaching out, very typical of dustbowl lands. The total impression to one coming from Maine which had been kept artificially green was that: "Desert Development is far along, near completion." We drove through a sandy, brownish-gray, dead-feeling landscape under a gray sky, depressed, but not disheartened.

Baltimore, Maryland, with its endless rows of identical brick houses, impressed us as especially oppressive. Driving in the car for several hours became sickening and irritating. It was after this a most dramatic change to pass into a zone of fresh, clear, blue energy, green foliage, cool breezes and fragrant smells approaching Washington, D. C. The clearness persisted throughout two days which I spent in the capital absorbing its meaningful design. Once more the impression was "someone is cloudbusting (deDORizing) here." An active change seemed to have taken place since OROP INFANT (June 1954) had first been carried out to demonstrate that relief from hot humid weather *is* possible. One could only conjecture that this pure atmosphere signified good news for Expedition Orop Desert Ea.

Washington was an "Oasis"; this became especially clear when we proceeded through Virginia towards the southwest. Twenty miles beyond the city the signs of desert development again appeared in full force: Signs of chronic drought appeared in the overgrazed pastures through which the underlying red loam and red clay showed; the low level of the streams in their beds, the parched, stunted look of corn stalks which had failed to develop fully; again, the trees bent and drooping. In Sperryville, Pa., I spoke to people in whose faces despair and listlessness reflected the desperate state of affairs in their environment. They knew about the severity of their situation: "Meadows and fields are burnt up, wells gone dry * * * people are sick, slowed down, dying * * *."

Approaching the eastern side of the Blue Ridge Mountains we noted that fields on the flanks of the hills were greener than those below in the valley. From the mountain-ridge at the "Skyline Drive" we saw for the first time the *"Desert Armor,"* in confirmation of the orgonomic desert theory. It was fresher on the ridge than in the valleys below. On the ridge, vegetation and trees looked sparkling, healthier, greener than down in the valley, similar to what is true of forested mountain-crests in deserts. Below the ridge, one could see the DOR-*layer* all formed, covering the earth to the distant horizon like a blanket, with a sharply delineated upper edge; beneath it the details of distant views were hidden in an opaque veil, as it were. The tops of distant mountain-ridges were seen to project clearly above the DOR shell, like islands above the ocean. This seemed to tell why forests survived in the high ranges (Sierras) while the lower vegetation, covered by the low-lying DOR blanket, died off, leaving only desert in the valleys.

MOUNTAIN PEAKS CLEAR ABOVE DOR BLANKET (*A. Blue sky; B. Mountain tops above* DOR; *C.* DOR *layer with sharp upper edge*).

Fig. 17

As the ridge road rose over peaks and dipped down into passes, one could subjectively feel the abrupt descent into the DOR layer: as a sudden pressure in head or chest, a sour taste in the mouth. One could also observe that while the trees sparkled and stood erect above the DOR ceiling, they drooped, were withered, and looked dark below it. The change occurred within a few yards sometimes. Below the DOR ceiling the rocks also showed more disintegration than above. I thought possibly a change in gravity function was involved in the transition between zones, and wondered whether some types of plane crashes might be due to such sudden zonal changes. I observed a northeastward bending of tree trunks, with Melanor and moss formation on the east side of trees.

As we descended finally into the Shenandoah Valley, what had looked like a blackish blanket over the ground now enveloped us from below as a blinding dry heat, with the mountains disappearing into a gray haze behind us. (It seemed that this layer contracted towards day's end, seen most clearly evenings and mornings, while it expanded during midday, under the sun.)

We were getting closer to the real desert:

"Kingsport, Va.: It is dry; lawns parched; a bright light on everything; droughty; note that the pall is darkest over towns * * *."

"Rogersville, Tennessee: Melanor reaction, rocks and earth black; slowed down movement of people * * *."

The dustbowl character of the land was clear 40 miles east of Knoxville, Tenn. Here in Tennessee everything was black: The burnt up cornstalks, the black telegraph poles, the black roadside gravel, the black tree stumps. The topsoil had become a dirty gray; black humus had dried out, turned to "dust". All this desperate area seemed to need was DOR-removal, water, plus an end to the atomic explosions (we were now 30 miles from Oak Ridge, Tenn.).

The patchy sandy areas of a beginning dustbowl were like islands of complete desert in a dying, but still alive landscape. Here there was no sparkle in spite of sunshine —a photograph taken in full daylight appears with the typical darkness of heavy DOR. The people appeared depressed, listless, quiet, slow-moving because of DOR. Clouds over Knoxville, Tenn., were steel-gray, fuzzy—of drought type. To our surprise there was a rather DOR-free zone to the *west* of Oak Ridge in which for the first time we saw sparkling, colorful autumn foliage. Red clay bluffs with contrasting green vegetation seemed typical of the healthier sections. Within these better looking stretches there now appeared pockets of desert: where whole stretches of forest were made up of skeleton-like ghost trees, and yellow, hard, cracked clay ground, rarefied fields and plantations, with only stubby, black, burnt crops remaining unharvested. Did the later completed desert patches correspond to these DOR pockets, and later oases to the reduced green areas? The land showed signs of overgrazing and erosion; the hillsides were gullied and the creekbeds dry. Driving through the hilly Tennessee country we noted repeatedly that the distribution of the DOR pockets had a relationship to the topography with regard to the west to east flow of OR energy: for each time the west side of hills appeared (1) slightly fresher, (2) with bluer haze, and (3) the vegetation greener and more alive than on the east side of ridges.

West of Wichita Falls, Texas, the drought situation again worsened, merging imperceptibly with semi-arid brushland and finally the deserts of New Mexico. Here we ran into the beginning of a sharp DOR zone, with DOR ceiling. Square, steel gray drought clouds were now distinct. Vegetation suffered, soils appeared eroded and baked. West of Seymour, Texas, we met the edge of the expanding desert. A waitress stated, "There has been drought for three years; the situation is desperate." The transition to desert was marked by a blackish DOR layer which extended low over the horizon in four to five stretched-out layers, the uppermost merging into a brighter but still blackish sky. The earth was sandy with dune-like formations. Whitish gravel covered everything. The DOR ceiling was located sharply towards W and SW, diminishing towards N, continuous towards S and disappearing in the E. The ceiling originated from several DOR layers which lay parallel over the plains. Desert characteristics increased as we proceeded westward. Red desert hills and terraces, mesas showing their tableshapes, with stratified red sandstone and clay rocks, dotted the plains. The highest layers were still gray, the granite crowning the mesas giving them a *"turret"*-like appearance, which posed a problem in itself.

Fig. 18

Pastures disappeared as we neared the New Mexico border at Bronco. Here a wide plain, covered with grayish white sand, blown by strong winds, stretching to the vast horizon completed the impression of "desert." Although it was very hot as we neared Roswell, New Mexico, no OR flow was visible on the road, which should have been "shimmering with 'heatwaves'." Instead, DOR was well marked to the west against purplish, black, barren mountains, in the sky as a blinding grayness, and over the horizon as a grayish layer. The caking of formerly good soil was progressively characteristic and eventually the caked soil prevailed over vegetation, which now consisted only of scattered low bushes, while grass disappeared.

After the desert valley it was a relief to spend a night at Ruidoso, New Mexico in the Sierra Blanca Mountains (near 7,000 feet). Here a strong, reactive, secondary vegetation had sprung up, again more marked on the western slopes.

Descending into the desert valley of Alamogordo, New Mexico, where the military White Sands Proving Grounds are located, we saw the plain covered towards W, SW, and N with a thick layer, several hundred feet high, of a gray, dead, opaque mass of DOR. Overhead the sky was blue-black, with some droughty, thin, high clouds. One felt a strongly salty taste. The white sand dunes showed a clear Orite accumulation. Could it be that White Sands was further attracting DOR? The DOR veil was the most remarkable we had yet seen, hanging thick and opaque, low over the landscape. The mountains edging this plain looked jagged, barren, with deep ravines "as if eaten out." About 20 miles beyond White Sands the air brightened, but DOR still prevailed. I remarked: "DOR *is eating up mountains, as it were.*" This spot was Sahara-like, without any vegetation.

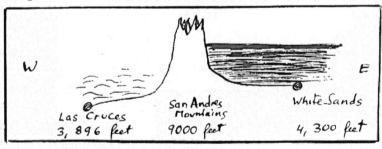

CROSS SECTION OF SAN ANDRES MOUNTAINS: DOR LAYER WITH CEILING ON EAST SIDE; LITTLE DOR ON WEST SIDE.

Fig. 19

We saw no sharply delineated DOR layer to the west of the San Andres mountains. In the pass the CPM ranged from 150 to 200, steady. To the west side the soil was redder, and a new type of prickly desert vegetation with yucca plants was seen.

I summarized the characteristic types of DESERT DOR as follows:

1. Heavy opaque low layer several hundred feet high, no sharp upper margin. Dense center attracting less dense periphery.

2. The DOR layer ceiling sharp towards outside, unsharp towards inner surface.

3. Purple-gray DOR in front of mountains.

4. Steel-blue DOR in front of mountains.

5. Gray dirty black discoloration of mountains.

On 10/29/1954 we arrived in Tucson, Arizona, which was to be the base of our Desert Ea operations. We found here a land different from what we had previously known: Here were flat plains, stretched out between mountain ridges, which rose abruptly in a general north-south direction. Contrary to expectation, the coloring was not brilliant, although the light had a blinding effect; rather all objects appeared grayish-white. This gray tint was noticeable especially in the sparse vegetation which spotted the gravelly plain. Life here had reacted with a *"secondary"* type vegetation to the presence of DOR: Here everything one touched was sharp, prickly, spine-covered. We were impressed by the *bare* ground, giving a general impression of whiteness, hardness. The surface was baked, with large clumps of stonelike crusts (caliche), found even on digging to a depth of several feet. The plains were traversed by sharply cut out dry streambeds, or "washes." These were evidence that the few rains which occur are torrential brief downpours; which, hitting the hard ground, simply run off, causing severe erosion of the bare ground. Later we were told that water had flowed in these riverbeds the year around as recently as 50 years ago. No prairie grass was to be seen anywhere.

The high evaporation rate due to the dehydrating effect of the DOR blanket thus presents a menace to life. The official relative humidity readings in Tucson when we arrived ranged between 5–10% at noon. At the end of January, 1955, the relative humidity was often between 60 and 90%. The DOR veil hung down as a gray-black blanket in front of the mountain ranges. The upper edge was usually well delineated, with mountaintops typically protruding out of the DOR ceiling like islands. Jet vapor trails either did not form at all, or held only a very brief time. Clouds were rare, thin, fuzzy, i.e., droughty.

Subjectively, we experienced the DOR atmosphere as oppressive, irritating; we felt drying out; we could hardly keep up by drinking fluids. Our mouths became parched, our voices hoarse, our lips cracked, our skins formed fine white scales. The blinding heat seemed to draw juice and life energy out of our bodies. It was an effort to walk a few hundred yards.

Thus having seen and felt the desert, Expedition Orop Desert Ea proceeded to begin Cloudbuster operations, in order to find out whether such a climate could be changed.

Survey of Travel Route

Day	Date	From	To	Miles	Daily Total
1	10-18-1954	Orgonon Rangeley, Maine.	Framingham Motor Court, near Boston, Mass.	288	288
2	10-20-1954	Boston, Mass.	Bo-Bet Motel, Mt. Ephraim, New Jersey.	288	576
3	10-21-1954	Mt. Ephraim, New Jersey.	Clarendon Hotel Court, Arlington, Virginia.	148	724
4	10-23-1954	Arlington, Virginia.	Lakeview Motel, Roanoke, Virginia.	250	974
5	10-24-1954	Roanoke, Virginia.	Mount Vernon Motel, Knoxville, Tennessee.	288	1262
6	10-25-1954	Knoxville, Tennessee.	Milan Plaza Motel, Milan, Tennessee.	320	1582
7	10-26-1954	Milan, Tennessee.	Park Plaza Motel, Texarkana, Arkansas.	396	1978
8	10-27-1954	Texarkana, Arkansas.	Siesta Motel, Seymour, Texas.	430	2408
9	10-28-1954	Seymour, Texas.	Nob Hill Lodge, Ruidoso, New Mexico.	412	2820
10	10-29-1954	Ruidoso, New Mexico.	Spanish Trail Motel, Tucson, Arizona.	396	3216

Transportation of the Cloudbuster and Laboratory Equipment from Rangeley, Maine, to Tucson, Arizona, October 7 to October 19, 1954

Compiled by: Robert A. McCullough

The following account summarizes notes made during the trip from Rangeley, Maine, to Tucson, Arizona, in which the WRF Ford truck, mounted with a 1953 model Cloudbuster and laboratory equipment, was taken to the U. S. Southwest desert area for Expedition Orop Desert Ea. The trip began on October 7th and terminated some 3,300 miles later on October 19th.

October 7, 1954:

Rangeley to Hopkinton, New Hampshire. The weather was clear, bright, sparkling and cold. There was no DOR. Concord, N. H., was free of DOR. Heavy frost at night—22° in the morning.

October 8, 1954:

Hopkinton, N. H., to Sangerfield, New York. There was a high thin overcast. From Keene, N. H., on the DOR was very heavy. Every town was cloaked with it, every valley was filled with it. At the height of land in Vermont, where one can see long vistas of mountains, all the valleys were filled with black DOR while the ridge tops were clear and sparkling.

First contact with chlorine treated water since coming to Orgonon in June, 1953. It wasn't wet—it left one's skin dry. After a shower with it, one felt scaley—dehydrated. It was not refreshing at all. It was as if the skin couldn't or wouldn't absorb it.

October 9, 1954:

From Sangerfield, New York, to just north of Erie, Pa. Heavy, thick overcast all day. All across upper New York state the DOR was heavy—whole valleys were just saturated with it. However, the trees did not show overt contraction and dying. The Finger Lakes were uniformly *green* in color.

October 10, 1954:

From Erie, Pa., to just west of Mansfield, Ohio. The weather varied from light showers to fairly heavy rains. (It was on this date when Chicago and northern Indiana had severe rain and resulting floods.) The DOR was very heavy. All the cities were "smoggy". In a conversation with a Clevelander, he commented that Cleveland didn't have winters any more—only smoggy drizzles * * * hardly any snow * * * miserable winters.

October 11, 1954:

From Mansfield, Ohio, to Pendleton, Indiana. The weather was interspersed with showers and clear between showers. The DOR was heavy, but twice following heavy showers the OR blue would come in beautifully from the west with a complete lack of DOR. Winds were strong southerly.

A possible cloudbuster effect must be mentioned here. While the cloudbuster was pointing downward and rearward, and was in addition stoppered and covered, it was noticed that clouds kept forming to the west of the truck. They would form, build and then, when overhead and rearward, they would dissipate. This was observed over several days and occurred too persistently to be a coincidence. A confirmatory observation was that this phenomenon was absent at night and twice when I stopped for several hours during the day.

October 12, 1954:

From Pendleton, Indiana, to just east of Hannibal, Missouri. Showers all morning, but largely cleared in the afternoon. The sky showed a lot of black DOR. Clear at night. Cold.

Comment on trees: After leaving Vermont, the only foliage change noted was a finely interspersed yellowing all along the route. There was no area of general yellowing and there were no reds or other colorings. It seemed that some individual trees in every grouping were contracting, leaving the yellow behind; but there was no bending, contraction from the terminals, or general dying of trees east of Kansas.

October 13, 1954:

From Hannibal, Missouri, to three miles east of Blair, Kansas, and then on to Hiawatha, Kansas. It was clear in the morning but soon N to S bands of clouds would form to the west (clear to the east and west of the clouds), thicken and then dissipate overhead and rearward. The DOR was heavy in the morning—especially on the horizons. It was not too bad directly overhead. It seemed oppressed to the ground and did not extend too far upward. Wind was SW all day.

There was a very severe local thunderstorm in the evening with Atchinson, Kansas, 20 miles to the south, reporting 2.17 inches in one hour.

October 14, 1954:

From Hiawatha, Kansas, to just west of Atwood, Kansas. A few clouds in the morning early which cleared. In the afternoon a north to south running line of unpredicted isolated showers developed to the west, but the winding of the road took us through and between them without rain. A *strong* north wind blew steadily all day. Water in ponds was blue after and just to the west of the showers. The DOR seemed to lie in bands. Driving west one would be in heavy DOR for some 20 miles and then there would be a DOR-free band of 20 to 60 miles and then DOR again. This was constant for the rest of the trip.

Atwood, Kansas, was very DORish. We stopped at a restaurant there for supper. It had fluorescent lights. Everything was dead in it: the waitresses looked and acted dead, the service was horrible, tempers all around were short. I had had a few "cold" symptoms the preceding morning: running eyes and nose, ear noises, sore throat, sneezing, etc. They had left after a few hours and had not reappeared. However, they came on again in that cafe and in 10 minutes they were full blown. I got out. I felt that if I had stayed in there 10 minutes longer I would have come down with double pneumonia. Our two year old boy *had* to get out also. He was all right outside. My wife reacted in strong shrinking. Another family nearby with two boys—aged about 10 and 12—ordered supper. The younger one wasn't hungry and the older one vomited his up. The father commented that he had never done that before. My "cold" symptoms left during the night.

October 15, 1954:

From Atwood, Kansas, to Kit Carson, Colorado. Clear, no clouds. The DOR lay in bands with areas of clear, sparkling blue sky in between. Vapor trails dissipated quickly in the DOR areas. From conversations all across the country, Saturday, October 2nd, seemed to be a day of general rain all over the U. S. It had rained and hailed 1½ inches at Cope, Colorado, and a heavy rain at Lamar. Also in New Mexico. The railroad town of Hugo, Colorado, seemed to be the one worst infested with DOR. It was putrid there.

The croplands showed desert development. There were corn fields where only half the crop had germinated and only half of that had reached a height of 8 to 12 inches. Milo and other sorghums have replaced wheat to a great extent. Very little wheat was seen in areas which formerly were solid wheat.

October 16, 1954:

From Kit Carson, Colorado, to Tucumcari, New Mexico. The weather was clear. What at first appeared to be vapor trails kept forming to the west as I drove south. It was finally determined that these were not vapor trails but actual clouds—long, thin, ropey, white clouds laying low to the west. They kept forming there *all day*. There was a tendency for some gathering in them. They would not form until I was almost opposite the area in which they did form. What apparently were two fires far to the south and southeast were observed in the day. After about 5 hours the smoke had all gathered into a big black cloud—immobile—over the area. It looked like a rain cloud.

Lamar, Colorado, had been the center of the dust bowl last spring. It was easy to see why. Lamar lies at the bottom of a long, deep depression in the plain. It is almost 1000 feet lower in altitude than towns 40 miles north and south of it. This depression would easily lend itself to stagnation. They had had some rain since the dust bowl and the soil was somewhat covered, but the scars remained. There were even sand dunes to the south of the town. In talking with residents of the town, it was found that the soil was very fertile—the only limiting factor being water. "With 20-25 inches of rain here a year we could grow anything." Dying trees were very evident in Lamar. The DOR increased from Eads south and was heaviest in Lamar. It lessened as I climbed to the south of Lamar.

Early in the morning I observed, at Eads, that the OR flow was more intense than I had ever seen it. The distortion of distant hills was terrific. As viewed through 10x glasses, the entire top of the hill would appear to be severed from the base, flow to the SE, and hang there. I was strongly reminded of both ocean waves and fire. The flow itself was W to E. It was a very moving sight.

Melanor was observed on rocky outcrops in western Oklahoma. In this area, also, Et* in the form of caliche and Orite was observed. Nowhere was Melanor especially prominent.

The soils of the milo fields in Texas north of Dalhart were examined, and they were found to be very light colored and powdery. In fact they were just dust. I am puzzled as to why they have not blown away entirely.

Right after I left Dalhart, Texas, I noticed that my neck had swollen grossly as well as did the parotid glands. They remained thus until the following morning.

October 17, 1954:

Drove from Tucumcari, New Mexico, to the White Sands National Monument, New Mexico. The OR flow all day was west to east. Clear. Tucumcari was free of DOR outside of the buildings. It was still free of DOR for 20 miles SW until one mesa, all alone, was observed to be enveloped in DOR. All the other mesas were red. This one was black as coal. As I drove further W and S the DOR increased until at Durand it was complete DOR. It remained thus until I came to Carrizozo. This area is high upland desert country covered with cedar (juniper) trees. This country has been very hard hit by desert development. A very high percentage of these trees are dead—have been for several years—and others are dead on their tops or in large areas of their sides. This is all red soil country. The land was very dusty and dried out. Evidences of a recent rain were seen, but the land had responded little.

---

* Et is a designation to be dealt with in a different context.

The DOR cleared rather sharply at Carrizozo. I visited the Mal Pais lava beds at Carrizozo. The whole valley here is covered with sheets of black rock. Only the surface of these rocks is black—deeper, under or inside it grades into a brownish sandstone type of matter. The surfaces are very pitted. I felt pressure in the area, but much less so than expected after experiences with Melanor in Maine. The black rocks themselves were much less noxious than expected.

The area between Carrizozo and Alamorgordo was DOR-free, and the distances were filled with a beautiful whitish-blue haze, OR blue. The DOR was concentrated heavily around Alamogordo and Holoman Air Force Base. It seemed to get even thicker there after dark. The town itself—a boom-town type—was very desolate.

I camped that night in the White Sands. There was no DOR there then. The shadows were blue. I felt no DOR— only one short period where my pulse speeded up a lot. However, there was something very moving about these many acres of white sandlike matter. One very noticeable behavior trait was noticed in myself as well as *all* others. This was the tendency to stand motionless for perhaps hours on the dune crests. One was just moved to do so. It was very noticeable.

There was no overt evidence of any lasting results from the first A-bomb blast which took place in this area.

October 18, 1954:

Drove from the White Sands area to 20 miles SE of Tucson, Arizona. The weather was clear and warm to hot. The DOR lay heavy in all valleys—seemingly heavier on the west sides than on the east sides of the valleys. The passes, mountain peaks and ridge tops were clear and free of DOR. New Mexico had much more DOR than did Arizona. The OR flow was west to east and was quite easily observed.

Fourteen years previously, I had traveled through this area and camped in it. It had been a lush verdant desert as deserts go—tall Joshua, cholla, yucca, mesquite, creosote bush and grasses. In traveling over it again all this was changed. There was just a plain on which grew a few short tumbleweeds and saltbrush—rarely reaching 12 inches high. All the vegetation was gone. Even the desert is dying. Again there were signs of recent rains in mudholes and playa lakes—but the land had not responded. The land was completely dried out—burned.

The picture changed from eastern Arizona westward. The range looked much better. Reports of natives told of good summer rains here.

October 19, 1954:

Drove into Tucson at 07:00 hrs. completing trip.

CHAPTER V
CHOICE OF TUCSON, ARIZONA

Tucson, Arizona, was chosen as the base for OROP Desert Ea for the following reasons:

1. It was situated in the southwestern corner of the U. S. A., only about 250-400 miles from the Pacific Ocean, at the southwestern entrance of the Galactic stream onto the continent.

2. The Tucson region was open in three directions: Toward southwest (the Pacific Ocean), north (Great Basin) and northeast (Great Plains). Moisture drawn from the Pacific toward the Tucson basin, would move freely in all these directions. It was later shown that moisture could also be drawn from Mexico to the south over the Santa Rita mountain range.

3. The region around Tuscon was reported as completely barren desert, i.e., with no prairie grass and no other primary vegetation growing in it. It was also reported as one of the hottest and oldest desert regions (25,000 years) of the U. S. A. It had not rained there for five years till 1954; the river beds had all been dry for about 50 years, according to reports on the spot.

4. One of the members of the advisory board to President Eisenhower established in 1953, Mr. Douglas, lived in Tucson and was head of the South Arizona Bank and Trust Company. I had hoped to meet him, since he had written requesting a personal discussion. For various accidental reasons, we never met. His banking institution helped along during the entire expedition in a most friendly and cooperative fashion.

[132]

The expedition was to prepare its equipment during September and to get under way sometime during October, 1954. During the summer a committee had been formed which was in charge of the handling and organization of all affairs of the Expedition. It consisted of the following men and women:

The *financial committee* consisted of Dr. Elsworth Baker, Fair Haven, New Jersey; Dr. Michael Silvert, New York; and William Steig, Cream Ridge, New Jersey. This committee was responsible for keeping the funds coming in at a rate of about 2000 dollars per month for running expenses. Altogether, the WR Foundation had a reserve of about 6000 dollars. In addition, I had designated personal royalties retained in Research Funds of two years, 1952 and 1953, 5400 dollars, to be spent for the purpose.

Orgonon as a home base had to be kept going on a minimum level. The Orur material was to be flown in from Orgonon, Maine, to Tucson, Arizona, later.

Also, the instruments and the archives at Orgonon had to be guarded, especially against intrusion by the drug agents who were at that time intent to break in to get information on our work for the chemical-atomic industry and their Russian political associates in their battle against the discovery of the life energy. The caretaker of Orgonon was in charge of this job. He was to keep in touch with Dr. Silvert in New York regarding the planned transport of the ORUR material. He had to guard this material which had been so distributed and hidden that no one could ever find it.

Baker and Steig were to collect funds in the East. Silvert was directed to establish a bookkeeping office for the expedition treasury and to stand by for the transportation of Orur over the 3200 miles to Tucson.

McCullough was ordered to go by truck with the equipment as the first party. He was instructed to report back to Orgonon until I departed, then to switch reporting to our social worker, Grethe Hoff in Boston, who would relay all messages during my transfer. This arrangement proved its value as will be shown later. We could not prove anything, but we all had the distinct impression that we were being watched by both the U. S. Air Force and possibly the Ea, to judge from certain happenings during the crossing of the continent and especially only a few hours after the arrival of the Orur material in Tucson on December 14, 1954.

McCullough carried the following equipment on his truck:

Laboratory equipment:

    1 SU-5 Tracerlab G-M survey meter
    1 Nuclear G-M meter
    1 Fisher laboratory scale: 0.1 to 610 gms.
    1 Stop watch
    1 Sextant in case
    1 Galvanometer, thermocouple, and 6 volt dry battery
    1 Sterilizer, 110 volts
    2 Small tripods
    1 Case of lab tools—syringes, spatulas, needle droppers, scissors, forceps, teasers, scrapers
    1 Coplin jar with approximately 100 microscope slides, assort.
    2 Wax pencils—1 red, 1 blue
    1 Bag of materials (WR's; compasses, etc.)
    1 Box assorted glassware—flasks, beakers, Bunsen burner, glass-working tip, graduated cylinders, Stendor dishes, separatory funnel, funnels, filter paper, side-arm flask, Buechner funnel, test tubes, ampules, vials, alcohol lamps, pipettes, cotton
    3 First aid kits

Laboratory equipment (continued):

1 Bottle live neutral Oe
1 Bottle of NaOH sticks
2 Table lights
1 Ice box
1 Telescope in case and tripod, 3.5 inch refractor
1 Right angle telescope
3 Wind vanes
1 Thermometer
1 Barometer
1 Relative humidity meter and tables
2 Typewriters, portable
1 Seal—WRF
1 Case of carpenter tools
2 Portable radios
3 Sets of charts and maps
2 CO_2 fire extinguishers
1 Tent and sleeping bags
1 Set field cooking utensils
1 OR shooter
1 OR funnel
1 OR blanket
1 Incinerator
1 Basket, lunch, wicker
1 Folding map table
1 Hair clipping unit

Photographic equipment

1 Leica 35 mm. camera
1 Sixtus light meter
1 Cine-Kodak Special movie camera
1 Telephoto lens
1 Bolsey 35 mm. camera and flash attachment
1 Weston light meter

Photographic equipment (continued):

   1 Tripod
   1 35 mm. slide projector
   1 Folding screen for slide projector
   1 case demonstration slides
   1 Kodascope 16 mm. projector

Microscopic equipment

   1 Reichert Z-microscope
      with 7x, 10x, 60x and 80x objectives
      with 5x and 16x oculars
   1 Microscope lamp and mount
   1 Transformer
   1 Monocular tube
   1 35 mm. Leica photographic adapter and two cables
   3 Spare pointed-filament bulbs
   1 Dissecting binoculars and stand
   1 Hand magnifying lens
   1 Large book of lens papers

William Moise, who had assisted me with most drawing operations since the summer of 1952, was to assist me with OROP Ea as operator, together with McCullough.

The fourth member of our staff was my daughter, Eva Reich, M.D. She was selected to be the physician of the expedition. She had given up her medical practice in Hancock, Maine. Her car carried equipment, personal belongings and was put at the disposal of the work for the entire period in a selfless manner, free of charge. She travelled with me in my Chrysler station wagon, took notes on observations during the journey and later compiled the notes in a publication "DOR Clouds over the USA" (CORE VII, 1-2). She became an acute observer of all matters connected with Ea, since she seemed to have developed a sharp sense of perception for atmospheric OR energy functions. Her scientific training, too, was of great help.

I cannot forego to mention the junior fifth member of my staff, my son, Ernest Peter Reich, at the time ten years old. He was most helpful in errands, and in other ways, up to full scale observations. It was he who made the discovery of a "node" while drawing from an Ea, to be reported later in detail in its proper context. Peter was special assistant to Moise in most drawing operations, when not in school.

CHAPTER VI

THE PLANETARY VALLEY FORGE

An Ea Attack on October 13, 1954

McCullough encountered an unusual storm on his route toward Arizona near St. Joseph, Missouri, in Kansas, on October 13, 1954, the same day we had suffered gravely under DOR at Orgonon. The following is his verbatim report, dispatched on October 19th from Arizona; it reached me via the Boston relay station by phone on October 22 at a motel in Washington, D. C. It was picked up by a listening post of the ATIC in a room beneath my motel room; it was hurried away by car immediately after the phone report was ended.

The McCullough report corroborated our previous experiences at Orgonon: The war of the planets was truly on. The Kansas event of October 13, 1954, was followed by further clear-cut Ea events in Arizona, especially on December 14th, 1954, a few hours only after the ORUR material was flown into the Tucson airport by Dr. Silvert. It was only during the preparation and integration of facts from the various log books that the seriousness of the interplanetary situation began to become clear in its manifold aspects. Until then, I had refused to accept the critical evidence without reservation. Here is McCullough's letter of October 19th, 1954, verbatim:

Received through G. H. from Boston to Washington, 10/22/54:

"Wilhelm Reich, M.D.
c/o Grethe Sharaf
25 Beaconsfield Road
Brookline, Mass.

October 19, 1954
General Delivery
Tucson, Arizona

Dear Dr. Reich:

I intend to type up a full account of my trip to Tucson but two incidents won't wait.

1. Wednesday, October 13. We had just left St. Joseph, Missouri in the late afternoon. Sky was darkening to the west. We stopped for an hour 3 miles E of Blair, Kansas. In the course of half an hour there was a tremendous storm buildup directly overhead. Altho it was dark, the landscape was continuously illuminated with cloud to cloud lightning. I suddenly was *sure* that Ea was building that up against me and the truck. Things happened just then. I saw a yellow moon-like Ea just appear below the cloud edge to the east. It disappeared immediately. Also observed by my daughter. Then a blinking white light was observed moving right to left in the *NW*. There was *NO* sound—and a plane would not fly into or under that storm. The feeling, the two sightings and the first rain all happened within 10 minutes—6:30 p.m. CST. We drove 20 miles to the west in terrific rain; I've never seen it so heavy. Finally got beyond it. The weather forecast had been for fair sunny weather—no showers. At 10 pm the Kansas City Weather Bureau forecast tornados SW → NE in that area of Kansas and Missouri. The nearest weather reporting station, Atchison, Kansas—20 miles on to the SW

of where we were—reported 2.37 inches of rain fell in less than 1 hour. I feel that this was an attack on the truck—??

2. No other Ea were observed until on the slopes of the mountains 20 miles SE of Tucson a yellow-orange dumbbell-shaped light was observed for about 30 seconds at 8 pm. At 12 midnight a B-47 stratojet bomber made a practice landing at Tucson, couldn't get up again and crashed and burned. One killed. Possibly a connection? (October 18, 1954).

DOR is very bad here. Eyes greatly irritated by it. Extremely dehydrating. It was never this bad in Yuma. Desert development has progressed a lot. Trees dying all along through Colorado and New Mexico. Even the deserts are dying. Where I camped in southern New Mexico 14 years ago in Yucca forests—only tumble weed is to be found now.

Sincerely yours,

/s/ Robert A. McCullough"

(Reproduction of original letter follows.)

Received through L.H.   October 19, 1954
from Boston & Washington   General Delivery
                               Tucson, Arizona

Wilhelm Reich, M.D.   10.22.1954
c/o Grethe Sharaf
25 Beaconsfield Road
Brookline, Mass.

Dear Dr. Reich:

    I intend to type up a full account of my trip to Tucson but two incidents won't wait.

    1. Wednesday October 13. We had just left St. Joseph Missouri in the late afternoon. Sky was darkening to the west. We stopped for an hour 3 miles E of Blair Kansas. In the course of half an hour there was a tremendous storm building directly overhead. Although it was dark — the landscape was continuously illuminated with cloud to cloud lightning. I suddenly was <u>sure</u> that Ea was building that up against me and the truck. Things happened just then. I saw a yellow — moon-like Ea just appear below the cloud edge to the east. It disappeared immediately. Also observed by my daughter. Then a blinking white light was observed moving right to left in the <u>NW</u>. There was <u>NO</u> sound — and a plane would not fly into or under that storm. The feeling, the two sightings and the first rain all happened

within 10 minutes – 6:30 pm CST. We drove 20 miles to the west in terrific rain. I've never seen it so heavy. Finally got beyond it. The weather forecast had been for fair sunny weather – no showers. At 10 pm the Kansas City Weather Bureau forecast tornados SW→NE in that area of Kansas & Missouri. The nearest weather reporting station Atchison Kansas – 20 miles or so to the SW of where we were reported 2.37 inches of rain fell in less than 1 hour. I feel that this was an attack on the truck ??

2. No other E↓ were observed until on the slopes of the mountains 20 miles SE of Tucson a yellow-orange dumbell shaped light was observed for about 30 seconds at 8 pm. At 12 midnight a B-47 Stratojet Bomber made a practice landing at Tucson, couldn't get up again and crashed & burned. One killed. Possibly a connection? (October 18, 1954)

DOR is very bad here. Eyes greatly irritated by it. Extremely dehydrating. It was never this bad in Yuma. Desert development has progressed a lot. Trees dying all along through Colorado and New Mexico. Even the deserts are dying. Where I camped in southern New Mexico 14 years ago in yucca forests – only tumble weed is to be found now.

Sincerely yours, Robert A. McCullough

The Tucson Base

With the help of The Southern Arizona Bank, a house with 50 acres land around, suitable for our purposes, was found and on October 31 we began establishing ourselves. The base was named *"Little Orgonon."* It was excellently located about eight miles north of Tucson, between route 80 and 89 to Oracle Junction and route 84 leading to the west.

Tucson itself is situated in a valley surrounded by interesting, at the time totally barren mountains. Only the crests of Mount Lemmon carry remnants of the old pine and spruce vegetation. Below a certain top level there is nothing but rock, gullies, steep abysses, secondary desert vegetation and dry river beds. The following sketch shows the approximate location of the surrounding mountain ranges.

a. Mountain ranges

*L. O. . . Little Orgonon.

b. Galactic stream

Fig. 20. Position of the Tucson base

These mountain ranges were splendid observation marks for OROP Desert Ea, among them particularly Mount Catalina which was situated to the northeast of Little Orgonon, slightly curved toward the southwest, i.e., toward the Pacific Ocean. This mountain range became our most important observation landmark. This range functioned as a reflector of OR energy and recipient of the incoming moisture. It was the first to show the greening process. Further to the north lay the Mt. Lemmon group of ranges. On Mt. Lemmon we drew in our high OR potential in order to accelerate the incoming of the moisture from the Pacific Ocean (see p. 164).

Many Ea were seen hanging in the sky during the nights of October 31 and November 1. The laboratory equipment was ready on November 2nd. Our actual DOR removal operations, too, began on November 2nd.

The first few weeks were devoted to exploratory trips in the region around Tucson. I tested the distribution of DOR, its mobility, the sensitivity of the atmosphere to drawing operations, the prevailing wind directions, the daily changes in moisture and temperature, the possibilities of cloud formation, etc. When we arrived and weeks thereafter, there were no clouds in the sky at all. We were told by local citizens that it had not rained in the Tucson region for more than five years before 1954.

Our base was equipped with an excellent observation deck. We put up the telescope, the star altitude meter, the hygrometer, a portable Geiger Counter and photographic equipment. The deck was equipped with chairs which made it easy for us to make sky observations in the lying down rather than turn-neck-back position.

DOR had been very bad during the past few nights. There was no doubt whatever about the DOR emergency in the desert. I soon noticed that DOR was less during the night, if there were no Ea in the sky; DOR usually increased during the day. DOR seemed to be responsible for the *burning* heat, the parching effects of the sun radiation.

My first impression of the surrounding mountain ranges was: *They were "eaten out," gnawed at by DOR as if a monster were feeding on mountain rocks.* This first impression was later confirmed by careful observation. The deep-cutting ravines and gullies impressed one as forcefully gnawed out and not as due to water erosion. There was no water whatever in the ravines or riverbeds. There had been no water for 50 years. There was no prairie grass whatever, only sand and gullies and caked, parched, sandy soil.

The desert gave the impression of being in flux. This impression was secured as final when within a period of six months I had travelled the route to Oracle and beyond, about 80 miles daily to and fro on routine observation. As I have mentioned earlier, getting thoroughly acquainted with the region under operation is of paramount importance second to none. A landscape has an expression and an emotional flavor like a human being or an animal. To learn to know this flavor and *to live with it* in good comfort takes time, patience, absence of prejudice and of arrogant know-it-all, or similar attitudes adverse to learning.

The gullies on the mountainsides, the turrets of granite and sand (see Fig. 18, p. 119), the sharp, rugged cliffs which seem to grow out of the valleys, sand dunes forming already where there is as yet no Sahara type of desert; the particular, parching dryness which matches the parching, beating down of the sun through blackish DOR ceilings, these are more than mere "sights" or "details." They are features of a particular kind of living being. I felt no particular sympathy or love for the desert. I like green pastures, moss-covered ground in birch or oak woods. But I could, after a few weeks' observation, understand the peculiar attachment which some people seem to have for the desert landscape and desert atmosphere. It is sympathy for a dying living thing that is still struggling for its life.

The theory of erosion as the source of ravines and amassing of silt has made it difficult to appreciate the DOR that gnaws at the rocks, parches the soil, turns loam into sand, dries out river beds, attacks mountain ranges and flattens them into rounded sand dunes.

The upper reaches of Mt. Lemmon were still covered with primary vegetation of evergreens and moss. It confirmed impressions we had had during the crossing of the continent: *The primal vegetation is still alive where DOR clouds have not reached the peaks of the mountains.* DOR tends to *sink down into the valleys* and to hover low over the landscape like a ceiling. Where DOR hovers, life dies rapidly. Life does not return, or it returns only as *secondary* vegetation, adapted in the chollas, the saguaros, the palo-verdes and various cactus forms in the southwestern U.S.A., to the DOR atmosphere.

Desert appeared after closer acquaintance to be a process rather than a static state of existence. It appeared as a continuous struggle of life against death, of dying and reaction against dying in various stages. To this, even the most desolate type of desert, the "Sahara" sand desert, like that near El Centro, California, did not seem to make an exception.

The following rough picture of the sequences in the development of deserts originated from the observations over the years since the DOR emergency had first struck at Orgonon in 1951:

Desert Development

In my present desert work the decay of trees in the forests of the U.S.A. occupies a central position. The basic problem is the death of vegetation, the development of desert and the nature of the agents responsible for this process. Comparative observations and experimentation in Maine, on the Atlantic coast, in the southwestern U.S.A. and in the laboratory leave little doubt as to the nature of the killer of Life.

The killer of Life is Dead Life Energy or DOR: in other words, the same primordial cosmic energy which creates, sustains and reproduces life under one definite *set of circumstances is the very killer of life when these conditions are absent.*

Moreover, Life Energy, which is massfree, primordial cosmic energy, creates its own "material" "*carriers,*" such as the indispensable organic building stones H, O, C, and N and their various compounds, H_2O, O_2, CO_2, carbohydrates, fats and proteins.

In the living organism, the Life Energy shows metabolism, a level of functioning ("Capacity level"), continuous creative functioning in producing red and white blood cells, and other constituents of membranous or nervous structures, in keeping the bio-chemistry of the organism integrated and well balanced: *Energy Economy,* "orderly energy household."

At the very basis of these life functions we find the dying of the *Life Energy itself;* the change from *OR* energy into so-called *DOR*, i.e., the *dead* Life Energy, comparable roughly to the remnants of burned coal. The analogy goes much further than indicated.

The process of dying of an organism seems, ultimately, to be no more than the dying of the Life Energy itself, the change from OR to DOR. With this basic change all secondary functions of life, energy metabolism, economy, integration of functions, cohesion of tissue, continuous reproduction of living matter and sap, also cease. In other words, the common functioning principle of all "Death" is immobilized Life Energy (cosmic primal energy) which for the lifetime of the special organism has directed its material carriers, the membranous structure and the life fluids, constantly in exchange and metabolizing with the Orgone Energy ocean outside, in the environment.

On our planet the basic carriers of Life Energy are water (H_2O) and Oxygen (O_2). No one can tell at present whether Life Energy produces different kinds of carriers on other heavenly bodies.

Certain conditions are required for the primal mass-free Life Energy to produce its carriers, water, Oxygen, Carbon, and Nitrogen and its further organic compounds.

Life Energy itself exists in space. It can be shown to exist in strong acids such as Aqua Regia, a mixture of Nitric and Hydrochloric acid. It also can be demonstrated at the pH of 9-12 in strong Hydroxide solutions. However, the carriers of Life Energy form membranous structures of life in a narrow range of pH 7-7.4 approximately only.

It appears as if Life Energy in organismic functioning of what we used to call "Life" were restricted to a kind of narrow existence on a razor's edge, as it were—the razor's edge being a neutral or approximately neutral narrow realm, characterized by 7-7.4 pH, between a wide killing realm of *acid* (H^+) on the one side (pH 1-6) and of *base* (OH^-) on the other side (pH 8-12).

So much is certain, on the basis of observation and experiment:

Life is at *present, under the given circumstances,* existing on the razor's edge between two kinds of deaths. How this will be *after* Life has become aware of itself and its way, we cannot tell. It is within the realm of possibilities, even of probabilities that Life will construct, create new, safer, broader ways of existence for itself with the knowledge of life it is about to acquire.

Life thus holds only a narrow wedge as its own domain in the infinite vastness of the cosmic energy. Organismic Life Energy metabolizes from and into the cosmic energy ocean. Respiration, feeding and direct radiation ("heat") are the basic vehicles of the metabolism of Life Energy between organism and environment. Within, the organism apparently metabolizes freshly taken in OR energy into DOR energy which is being expelled in the form of CO_2, urine (NH_3 products), feces, sweat and gaseous exudations.

Energy equilibrium between charge and discharge is easily maintained in the healthy organism. During sickness more OR seems to change into DOR, also more DOR seems to be retained in the tissues. Thus, a prevalence of DOR energy would be a *basic* feature of all disease. From this prevalence several consequences may legitimately be derived.

One is, a greater inclination to exudate and to retain fluid in cavities, within the interstitia of tissue, in the abdomen, extremities in the form of edema, ascites, parenchymatous swelling, etc. There seems to be more to this inclination to retain fluid than mere inhibition of mechanical movements of body fluids. There seems to exist a close relation of DOR to water or rather to a change of DOR into H_2O. In alcohol addicts we find parched lips and tongue, thirst and severe dehydration of tissues with the direction toward cirrhosis of liver, induration of brain tissue, etc. DOR prevalence means need for more fluid, foremost water. OR energy shows, as we well know, great mutual affinity to water. The water is being absorbed wherever it can be obtained by the thirsty DOR energy. If we add to the great absorption of water the *change into water*, we obtain a truer picture of the *dehydration* of whatever is near and contains fluid. With the dehydration of tissues goes a lesser ability to move and metabolize fluids. Polyuria, uremia, albumin and sugar in the urine as signs of tissue disintegration, are clear results of a disturbed energy metabolism. Much detail remains to be gathered here; but the basic disturbance is clear: Dehydration of tissues, stagnation of metabolism, to which later is added lack of oxygen intake and the resultant CO_2 surplus as in cancer, in edema of the lungs, lividity, and cyanosis, recession of vitality on the whole. So far, the DOR manifestations in the organism. In a vicious circle, inhibition of OR functions leads to prevalence of DOR functions, and the latter lead to steady increase in OR function decay, thus to death.

DOR functions are characterized by a silent, invisible and inaudible, as it were, gnawing away and insidious consumption of the life force of a host or organism. DOR works like a tapeworm, within the intestines of the host, be it emotionally or substantially. DOR is hungry for nourishment, for water, for oxygen, for * * * becoming productive Life Energy, or OR energy, again, in short for "revival." But, exactly in its attempt to become fully functioning life again, DOR destroys its own host, its own source of nourishment, its own hope.

If emotional pestilent reactions are due to DOR, then the emotional desert does exactly to its host or giver what the tapeworm or DOR in the tissues does: Kill the host, the giver of life in silent destruction by way of sapping its strength.

This fact has obviously clear bearing on present-day social conditions:

Where life has been lacking nourishment and care for ages, as in the great Asiatic communities, DOR energy is prevalent, both in the living beings and in the atmosphere. The craving for nourishment is immense. The prevalence of DOR will attract more DOR, emotionally and physically. Complete destruction of the host, in this case of the globe of mother earth, looms on the horizon of our future.

It has great significance for the mastery of our future, that DOR surplus causes deserts in the landscape as it does in the organism. Desert souls will enhance desert development; and desert development will increase DOR or staleness in the human emotions.

The outcome hinges clearly on whether at all, to what extent and at what step of the decay process, DOR energy can be reverted again into OR or Life Energy. *The process of disintegration is clearly and doubtlessly reversible.* This has been shown in the desert work 1954/1955 in Arizona, when completely barren desert land was turned into green pastures again after thousands of years through removal of DOR from the atmosphere.

The process of disintegration of trees and whole forests is due to progressive DOR prevalence in the atmosphere. A slight DOR prevalence causes dryness, dryness in turn increases DOR. Thus, in a vicious circle, the water-hunger grows together with diminishing precipitation. The process is slow, and not easily discernible. Not much is known about its secret attrition of life. When the atmosphere becomes droughty, causing clouds increasingly to lose their cohesion and to dissipate more easily, the level of atmospheric moisture sinks and the land begins to parch. Water levels, too, sink. The Life Energy in the atmosphere has less and less resources of moisture. In addition, DOR causes the sun heat to turn into "burning" or "parching" heat. Thus the Life Energy turns, thirsty to the extreme, toward the moisture present within the vegetation. DOR penetrates the trees slowly from the top downward, and from the bark inward.

The atmospheric (or cosmic) source of the destructive DOR, effecting desert conditions, was established beyond doubt in 1953. The atmospheric DOR clouds exerted an effect of gradual immobilization of life in plants and animals. Concentrated Sodium Hydroxide (NaOH) in open dishes causes a white substance, termed *"white ORENE"*

(Oe) to settle down *above* the level of the fluid on the inner wall of the dish. As long as there is fluid in the dish, Oe remains moist and harmless. Under the microscope it displays beautifully formed "bags" that grow, segment, expand, and are productive in many other bio-energetic ways.

On the other hand, however, when the fluid is not replenished and the dish becomes dry, white Orene turns into a hard, white substance. It is *dead,* material Life Energy, Lt: "t" denoting the function of the deadly T-bodies, discovered 1936 in my Oslo Laboratory in a culture from sarcoma tissue. The hardened white Lt is true DOR substance. It irritates the atmosphere; it causes inflammation on mucous membranes; it results in T-bodies when kept, after full hardening, in water. There are other important qualities.

One of the most important ones is the turning black of "white Orene," viewed microscopically and, the other way around, turning white of Melanor *(Me), the black* matter-like substance which can be collected from the atmosphere. One laboratory research assistant had grown white Orene easily in her home in a Washington suburb, but failed to obtain it in her laboratory at N.I.H. (National Institute of Health). There, apparently the fluorescent lights had killed the Le outright and drawn it into the inside of the Argon tubes. One may easily see what a "control experiment" designed to refute the Orgone theory by all means, conducted by any prejudiced bio-chemist under such conditions could do to disavow the discovery of the Life Energy. (See also "Melanor, Orite, Brownite and Orene, Preliminary Chemical Analysis," in CORE, Vol. VII, Nos. 1-2, 1955, Orgone Institute Press, N. Y.)

This change of color is probably related to changes in the direction of radiation. In agreement with *Kirchhoff's* finding, the "black" Orene *absorbs* energy, while the *white* Orene gives off energy. This may gain some significance at later investigations. Whatever the reason may be, we may observe the same changes when Melanor from the upper atmosphere attacks trees in a forest. The attack is due to the droughty lack of moisture in the atmosphere. The moisture is now being obtained from the bark at first, then, upon further penetration of the tree structure, also from the deeper rings or layers. Correspondingly we see at first the bark getting blackish; then the bark disintegrates, and disappears. The process never sets in from the roots upward; it is thus not due to "bugs." The disappearance of the bark regularly begins at the tree tops, working its way downward toward the roots. Also, the blackening and ensuing disintegration of the bark begins on the upper sides of the branches; this points again clearly to the atmosphere as the source of the noxious agent, the Melanor.

After some time, when the bark is well off the attacked tree and the disintegration fairly well advanced toward the roots, the inside of the tree substance disappears, too: The tree becomes hollow. Then, due to loss of energy and substance, the tree bends or curls up like a cork screw; the branches sink down and fall off, until the whole tree collapses. Thus, Melanor (DOR) has robbed the tree of its moisture, of its alive OR energy, of its substance. Now we see the surface of the barkless stem and branches turning white. Melanor has changed into white Orene, exactly the opposite of alive, white Orene turning black and hardening upon loss of moisture. These processes can be directly observed on a far larger scale in the Sahara-like desert of the Southwestern U. S. A. near Yuma. Large stretches of "Black Rock," i.e., Melanor, are impressive as witness to the deadly force of DOR. This is true for both the outer Desert like the Sahara and the Emotional Desert as presented together in the excellent film, *"Bad Day at* BLACK ROCK,*"* 1955, with Spencer Tracy in the master role.

These observations point toward crucial processes of death and desert development at the roots of life. These processes are, to repeat, characterized by their silence, and by slow attrition of the victim. The life force having been sapped from the victim, it slowly becomes paralyzed and finally gives up.

The pestilential character shows the same type of behavior. He feels "black" inside and often is actually black at the skin. He saps juicy, emotionally rich people, deprives them of their strength, akin to the behavior of a tapeworm, within the host victim. The pestilential character thrives on the energy loss in the victim, but in the end he perishes with the host. From here to sociological conclusions regarding the secret dynamics of political dictatorship is only a logical step; from here, too, a bridge can be built toward understanding the connection between desert development on our planet and visitors from outer space. These visitors are using fresh cosmic energy for their locomotion and pour the slag, DOR, into our atmosphere. Whether this is being done on purpose or by accident does not matter as far as the effects upon life on earth are concerned.

Thus, to summarize, orgonomic research has found at the very roots of existence an energy, which, dependent on circumstances, functions either as a life giving, life furthering, and reproducing force or, in the absence of such conditions, turns into a killer of Life. It even becomes a destroyer of lifeless matter, as was shown in the disintegration of solid rock, granite, etc., to desert sand during the Oranur Experiment, 1951-1954.

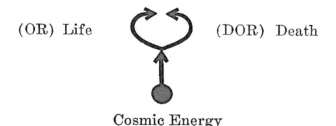

(OR) Life   (DOR) Death

Cosmic Energy

Phases

I. DOR ATTACK UPON GREEN LANDS

Original vegetation dies slowly due to repeated, severe droughts. It decays gradually, thins out, but at times it reacts with great lushness and abundant growth as if feverishly reacting against the DOR attack. Trees decay; they disintegrate in the center bionously; large and ever larger hollows develop in the trunks; then they bend like rubber hoses; slowly branch by branch, twig by twig they die until the stem topples over, too. Armored man passes by this devastation over the ages without noticing it. Or if he notices the insidious disaster, he tells nothing to anyone. He is slowly deteriorating himself, as clearly seen in the difference between an inhabitant of the Nile Basin or the Italian Peninsula only two and a half thousand years ago and at present in the 20th century.

In this first phase, water wells dry out, one by one, over years of drought and scanty rain. The surface soil cakes and turns white due to infiltrating Orite. The soil still has the power to come back upon a good, soaking rain or upon relief from the nauseating, dehydrating, life-draining DOR clouds. The caked soil, formerly black and rich and juicy, turns into loam, then yellow clay, and gradually cakes into little hard pebbles, not distinguishable from small rocks. An open eye and a functional mind, following processes in nature as they are, can see all this as an ever-changing function up till the Sahara-type devastation. In this phase, granite disintegrates imperceptibly, unless the eye is trained to see its progress; but the rock disintegrating into bions on the surface is still capable of reorganization, usually into a harder surface with a whitish hue, the unmistakable sign of Orite mixed into the bionous mass.

II. DOR ATTACK UPON CAKED SOIL

The caked soil becomes fissured; it does not come back as easily as before after a soaking rain. DOR has eaten its way into narrow and shallow fissures, deepening and widening them. The soil as well as the decaying rocks turns into gray, later *red clay*. "RED SANDS" begin to show. Loam, still capable of bearing fruit, develops from the clay through Melanor.

III. DOR ATTACK UPON RED CLAY

The caked clay turns more and more into *sand*. DOR destroys the cohesive power of the energy which kept the clay particles together and tends back toward rock formation on a smaller scale, as it were. This dry sand is, as shown easily by experiments, still capable of being reversed by moisture toward soil. However, if DOR continues to penetrate, to dehydrate, to cause everything in its path to disintegrate, the limit of reversibility is passed, and the sand becomes fruitless, sterile, unchangeable by natural means. Nature, then, still continues to create but no longer on the level of the living.

In this third phase, when the resistance of the living bas been broken by DOR and Melanor, when there is no OR life left in the soil, *mountains are literally being eaten up* by DOR. *MELANOR EATS MOUNTAINS.* This was the most dramatic experience in the desert, especially in the region around El Centro: Melanor eats mountains.

The mountains in the desert are being levelled off with single untouched rocks standing upright like turrets surrounded by caked clay and sand. Where there is still some moisture available, secondary vegetation is still possible. But the secondary desert vegetation will soon die, too. The rivers have died long ago. Deep wells are still there but they are losing out, too. The deep water well level sinks from year to year by a few feet. This was the case over the last decade in Arizona, according to reports of farmers. With the general decay, the hills flatten and round out. The landscape, due to ORITE deposits, acquires more and more a *whitish* look. The next step is:

IV. DOR ATTACK UPON THE CLAY AND THE REMNANTS OF LIFE

The *"WHITE SANDS"* only, sky steel gray above, hot radiation, Orite taking over, changing the desert into a whitish moon landscape.

This was the picture before me on November 2, 1954.

Proto-Vegetation — The Greening of Sandy Desert

It was not the primary objective of the expedition to "make rain over rainless desert lands." It turned out later that "making rain," if we could have done it, would have hidden from us the available scientific information on the dynamics of desert functioning. On the contrary, it helped greatly that rain could not easily be obtained. We had no ambition to impress anyone with making rain over desert.

During the first days of November 1954 the surrounding desert region, particularly at Mt. Catalina, *reacted to the systematic DOR removal with greening.* Shoots of prairie grass sprouted first singly, then in patches and finally the sand was covered with a fine carpet of green. The green spread toward Mt. Catalina, climbed during the following weeks up the mountain slopes, extended slowly toward the north along the highway; from here it spread out toward east and west, until by December last it stood several inches to a foot deep over a territory of about 40 to 80 miles from Tucson with prevalence to the east and north. I repeat: *Grass grew knee deep on a territory where no grass had been before,* where only barren sand had been as far back as people remembered.

The greening of the parched land came about *without a drop of rain*. The moisture we had been drawing in from the southwestern shores of the U.S.A. had accumulated in the atmosphere, was reflected by the slopes of Mt. Catalina, and had caused the greening of the desert lands. Now, such a phenomenon is incomprehensible to a mechanistic, "spore"-bound biology. According to the spore hypothesis the seeds of the prairie grass blades must have slumbered in the parched, caked, completely barren sand for thousands of years. Then, suddenly, with no rain falling nor soaking the ground, the shoots began to appear, as if from nowhere.

It is exactly at such crossroads of natural events that the failure of mechanistic thinking regarding fundamental matters of nature becomes evident. It is also here where orgonomic biophysics passes its tests. The mechanistic theory of spores was inapplicable. Still, the greening was a fact, observed by many people in that region. Cattle were driven, first by twos and fours, later in herds upon the new prairie lands. The mechanistic biologist kept quiet. He had nothing to say. Where his understanding failed and when his pride was hurt, he was inclined to simply misconstrue opinion in a rather unpleasant fashion, painful to listen to.

In order to obtain understanding of the appearance of prairie grass without rain, I set up a few dishes with caked soil for microscopic observation. One dish contained nothing but dry soil taken from the yard; it was kept exposed to the burning sun. Another dish contained soil which had been sprinkled with white Orite, Et so-called. It, too, was kept in the sun. A third dish contained caked soil taken from the yard, filled with well water to imitate the condition of irrigation. A fourth dish contained the same soil, but did not contain any water together with the soil; rather the water was arranged in such a manner that it *evaporated only* into the caked soil; this *arrangement was* designed to reproduce the actual situation of grass growing without rain. It was clearly the moisture increase in the atmosphere which had caused the grass to grow.

The four dishes were observed at regular intervals with about 300 and 1000x magnification over a period of several months. Soon several riddles were no riddles any longer.

The caked soil kept dry, did not change much, except becoming somewhat harder. The caked soil where Et had been added caked more and faster; after two months there were small pebbles of hard rock forming in the dish. The dish which contained soil plus water showed only bionous development with some protozoa.

However, the surprise came from the fourth dish. The soil had absorbed moisture from the surrounding water directly. Microscopic examination revealed an unknown type of growth: *Fan-like formations at the margins of bion heaps.*

Fig. 21. Dry sand (caked soil particles) to left on slide develop finger-like formations stretching toward water droplets to right on slide. Actual appearance at appr. 600-1000x magnification.

These formations could be reproduced at will. A few grains of caked soil are put on a clean, flat slide. Two drops of water are deposited one to each side of the small heap of sand. Observation over an hour or so with 300 to 600x magnification shows clearly *Proto-vegetation.*

The soil particles stretch out toward the moisture by forming digit-like protrusions in the direction of the droplets. This solved the riddle of prairie grass growing on parched lands without rain, upon DOR removal and increase in atmospheric moisture alone.

PROTO-VEGETATION must have been the original form of primordial vegetation on earth. Before spores or seeds could be there, some living form that *produced* seeds and spores must have been formed. The seeds could not possibly have come from nowhere; nor could they have passed through cosmic spaces with near absolute freezing temperature. We are witnessing the writing of an entirely new chapter of biogenesis on this planet and shall do well to stop at this point before proceeding again.

We understand now why rain would have obliterated this finding. It would have kept the Proto-vegetation from our view and, in addition, it would have drowned the outstretching bions as was shown by actual experiment. In a similar vein, also, man dying of thirst must first get used to small amounts of water. He must adapt himself to it. This finding corresponds in the manner it was secured to the finding of the star track which was longer and deviating at the same time. It is in finding such gems of knowledge that we also find our greatest satisfaction in knowledge.

Ea and Rain Clouds

On November 7th the moisture had risen in the atmosphere from the usual 15% to 65% R.H., an unheard of Relative Humidity for Tucson. A Tucson radio announcer who reported on the humidity that day also announced that he had been "told to announce only minimal humidities."

Some of the chollas and saguaros higher up on the plateau toward Oracle were blackening and falling apart where the green prairie grass had appeared. The rapidity of the reactions was astonishing.

We were drawing now with the Cloudbuster continually, from the southwest mostly. The task now was to obtain a continuous flow of moisture from either the western or the southwestern Pacific coast, through Yuma or Mexico, or if possible from both. DOR lay heavy to the west, constituting a barrier, as it were, against unimpeded flow of moisture. It was further contemplated to accomplish, if at all possible, a fusion of the western or southwestern moisture flow with the Atlantic moisture. But we knew well that the deadlock in the Great Basin due to the DOR ceiling had to be broken first. A first wedge had been driven into the staleness of DOR.

On November 7th the first clouds were forming thickly over Little Orgonon. They soon covered all of the sky tending toward rain. The OR flow that day was from east to west. Clouds were reported forming at the Pacific Ocean. At 13:00 hrs. the sky was completely overclouded, especially to the west. At 17:00 the sky in zenith was droughty. At 21:00 the clouds were dissolving rapidly, in a most conspicuous fashion. At 22:00 the sky was clear of clouds. We did not understand what had caused the dissolution of the rain clouds. We had, without being aware of it at the moment, come into contact with the effects of Ea upon cloud formation in deserts:

All through the night on November 7th to 8th, 1954, we witnessed a spectacle which opened up the realm of Ea with respect to cloud and rain formation. That evening a bright, large luminating ball was seen coming up like a star on the northern shoulder of Mt. Catalina. It moved very slowly toward the south along the mountain crest. Before reaching the southern slope, it turned upward and remained hanging about 10-15 degrees in one place for hours. The day had been replete with severe DOR; blankets of blackness hovered over the Tucson region and Little Orgonon. We drew off some DOR early that day, but it returned soon. The mountains to the south and west were especially and severely veiled with DOR blankets.

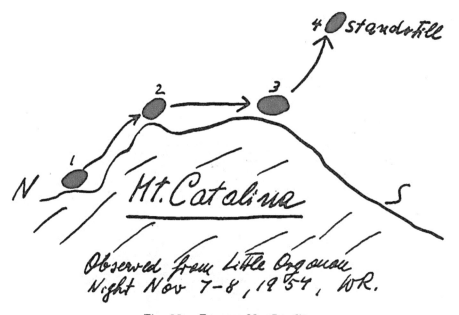

Fig. 22. Ea over Mt. Catalina

There could be little doubt as to the connection between the observation during the night of the bright luminating ball and the disappearance of the rain clouds on the 7th as well as the severe DOR clouds the following day, November 8th. The Ea had not been there before; it was not there on the following nights. There was no escape from the fact that we were at war with a power unknown to man on earth.

In addition to the grave task confronting us, the meteorologists in Tucson seemed to have concluded that, speaking in terms of publicity, we should be obliterated. All reports of the weather bureau were somehow slanted against our facts; the rise in moisture was not reported; the DOR blankets were ignored; the cloud formations were not mentioned or were talked away by some evasive comment; there was outright falsification of fact. We soon gave up listening to these dishonest reports. The situation somewhat improved as the weeks passed by. The coming-in of moisture from the Pacific Ocean was mentioned in the reports. The reports were at least trying to be fair. Something had changed in the Weather Bureau in Tucson. There was, at last, some respect for fact noticeable in the reports. But the sabotage of our efforts continued as we shall learn later on.

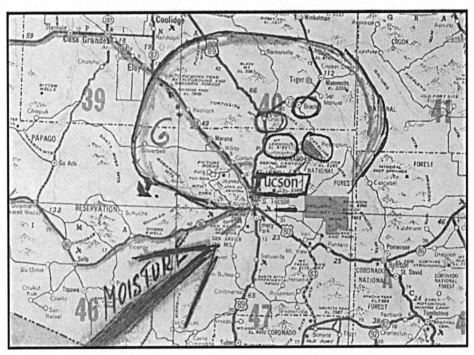

Fig. 23. The greening experimental area at Tucson

CHAPTER VII

THE Ea BATTLE OF TUCSON
Ea, DOR and Cloud Formation

The following theoretical work positions were held November 10, 1954:

1. DOR is the basic drought-causing and desert-supporting factor. There is some obscure relationship of DOR to water. DOR *"eats mountains and levels them off to sand dunes."*

2. Sahara sand is caked, crystallized soil mixed with Orite.

3. Ea causes strong DOR and dissolves clouds, prevents cloud formation.

4. Ea can be weakened or even extinguished by drawing off energy with the Spacegun.

5. Ea has caused the deserts of the planet, supported by earthman's emotional desert.

The days and nights from November 6 to 9 were filled with DOR sickness, dispersal of artificially created clouds, appearance at irregular times and places of one to two Ea in the East. There was no DOR accumulation after successful drawing was performed on the evening of November 8th from the eastern Ea. The morning of November 9th was free of DOR. The mountains were clear, the valleys brilliant. Such changes from DOR to DOR-free days and vice versa, which followed our drawing operations, were most convincing. So was, too, heavy DOR following sighting of Ea or not drawing from Ea. Thus, our position gained ground regarding the interrelation of DOR-Drought and DOR-Ea. We also learned step by step, to integrate malignant human reactions with outer, atmospheric DOR situations. *"You are dorized today,"* became a standing phrase to describe a behavior of confusion, crisscross activities ("pranking") and discomfort, clearly due to DOR.

The strong cloud formation of November 9th had been brought about by the night drawing operation of the 8th. This success grew further into the atmospheric events of November 10th. The effects of the Ea operations covered the region of Mexico and southern California, a region plagued with drought for years.

At 10 a.m. on November 10th a strong southwest wind arose. We all felt and saw moisture coming in heavy from the ocean. A bank of rain clouds hung low to the west and south. At 11 a.m., lasting till about noon, the wind shifted to east and southeast. The mountain ranges south of Tucson toward Mexico were submerged in cloudy mists. We phoned the Weather Bureau; they told us that all Mexico and the coast up to California were in clouds, 30,000 feet high. The draw from the east during the night certainly had produced strong results. We kept drawing from the zenith in order to keep DOR out as much as possible, also to keep the passage free from DOR for the clouds, should they pass overhead.

The cloud bank grew steadily in height and width. The wind blew strongly from the east at 10 miles per hour with gusts up to 20 miles. The wind blew strongest in Tucson where DOR was always heavy. I understood this detail only months later when I found out that gusty winds are due to OR hunting DOR ahead of itself. Dust devils and tornadoes became understandable as results of a *self-cleaning process* in the atmosphere, with OR cleaning out DOR.

At noon, *finger-like extensions* of the clouds formed overhead, reaching out far ahead of the cloud bank toward the north. The official weather reports registered rain in Los Angeles, one inch, at Mt. Wilson, one inch. The clouds held heavy from San Diego to Pt. Conception. All southern Utah was reported as covered with clouds. No rain was predicted for Arizona but it was raining at 13:30 hrs. in western Arizona. Showers were expected in Tucson during the coming night. It rained heavy all over the west coast, strongest in the southern parts on November 11th.

However, *no rain came to the Tucson region that day, contrary to prediction.* But a heavy cloud bank to the southwest, west and northwest along the horizon was preceded by a dense bank of fog, several hundred feet high. The moisture was strongly felt in the air. The east to west draw had built a high potential in the west which drew moisture streaming in from the ocean. This moisture was picked up by the Galactic OR stream and was carried toward the northeast. There had been much rain and snow in the northeastern U.S.A. during the winter months of 1954-1955. (See Fig. 20b, p. 144.)

All these events were ignored by the Institute for Atmospheric Physics at Arizona University. They were busy with a new project, counting the droplets condensing around dust particles per volume of air. They had received $150,000 from the National Academy of Sciences in Washington. I had applied for a similar amount a few months earlier, without success. I do not know what came of the counting of droplets in the air.

To continue: I assumed that day, erroneously, that the east to west drawing had impeded the incoming of the rain from the west. Later, I thought that an obstacle must have existed to the west which prevented the moisture from streaming into the Tucson basin. The moisture also seemed to pass over the Tucson region without forming durable clouds. It streamed northward as if drawn in strongly by the desert regions in the Utah area. The DOR over the northern parts of the desert basin may well have had this attracting influence upon the direction of the moisture flow. My original choice of Tucson as an open crossroad for three or four directions of flow seemed confirmed by the results of actual moisture flow. However, there was trouble ahead, since I knew at that time nothing of the *"BARRIER"* that barred transition of clouds to the east at the mountain range west of El Centro. I had to combat it March, 1955, before I could obtain rain over the American Sahara.

DOR thus was found to do two things: *It drew moisture to itself, due to its water hunger, and at the same time it prevented the condensation of that same moisture into rain clouds.* Much was in the dark here, of course.

The Tucson basin, the hottest spot in the U.S. southwestern desert, may for 25,000 years have been submitted to Ea attacks without man having been aware of it. Were the Ea which we saw in the sky possibly space machines which had been keeping deserts going for ages, preventing rain all along the times? No one could tell. But it was very much within the limits of reasonable possibilities. It could not and should not be ignored.

It rained on November 12th in Los Angeles, the drought having been broken there by 1.3 inches of rain; in San Diego with 0.9 in., Nevada 0.8 in., Grand Canyon 0.4 in., and Yuma 0.01 in. The humidity had risen at our base between 11/4 and 11/12 from 5 to 10% to 55% at noontime. The atmospheric background CPM had fallen from about 200 to 20 during the same period.

On November 13th the relative humidity went up to 67% at noon. Rain clouds lay heavy and thick on the mountain-tops. Greening proceeded apace on all mountain ranges surrounding our base. Yellow prairie grass was showing. It spread rapidly.

There was no doubt: *Slow increase of moisture in the atmosphere was to be preferred by far to a sudden rain over the desert. The soil had to have the opportunity to soak up the moisture from the atmosphere at its own pace according to need.* The moisture in the soil could be felt directly by touching it. The sand became darker and glittered from dew in the morning. It was on November 13th that it became clear to me why high mountain-tops in deserts are covered with vegetation while in the valleys there is no longer anything left of the primary growth: DOR sinking heavy and low into the valleys would gorge up all moisture quickly, while the DOR-free peaks would hold their water and water vapor. Also, the clouds would pour their moisture around mountain-tops rather than have enough time to accumulate in the valleys. It is well known that the sudden bursts of heavy rain and the run-offs have little value as far as soil preservation and farming is concerned. The water is not being absorbed by the soil since the soil is *not capable of absorbing it. It has first to adjust to the moisture, to seep it up gradually and to change slowly its inner, microscopic structure in turning back from caked pebble to bionous soil.* These processes can be observed with little effort if these principles are known and are used in the setting up of appropriate observational and experimental settings. This requires, of course, that one frees oneself completely from many wrong assumptions we encounter in the literature on the subject regarding "spores," "gullies being due to washout erosion," etc. *Drought precedes the formation of gullies.*

On November 13th the relative humidity at 17:00 was 45% as against 5 or 10% only 10 days before in the late afternoon.

On November 14th two bright, pulsating, flashing Ea were in the sky low to the east again, one approximately 15, the other 25 degrees up. Upon drawing, the first dimmed after an initial stronger blinking and remained dimmed. The second Ea wobbled first for a while, then it, too, dimmed strongly. Suddenly, a third suspect came up in the east, appearing suddenly as if from nowhere. Sudden appearance and disappearance are basic characteristics of Ea, well known to us. (See Ea appearing suddenly, photo 3B, p. 21.)

All through the latter half of November we were aware of the social events which intermingled with our Ea operations. Somehow, the two realms, seemingly so far apart, had a close bearing on each other. People in all walks of life were dimly aware of the cosmic functions affecting our planet; there was silent, but eloquent interest in our operations from farmers, laborers, in the bank, from government, and especially from those who tried to destroy us.

On November 18th, Judge Clifford in Portland, Maine, denied the intervention of the 15 physicians against the unlawful injunction. All physicians were, contrary to the original document, now exempted from the terms of the unconstitutional order. Only WR "Ad personam" was chosen to be the victim. The U. S. drug Hig later dragged Dr. Silvert back into the mud. Two drug Higs tried to get to Orgonon, but were sent away by the caretaker. Our literature continued to be distributed and the money came in regularly. Dr. Baker in New Jersey was highly instrumental in keeping things going.

On November 22, the United Nations Council accepted Eisenhower's "Atoms for Peace" proposal.

On the 19th, vapor trails, written into the sky by jets, were holding together, demonstrating an extension of the DOR-free region both upward and outward beyond Mt. Catalina. But the Ea were not inactive either. On the 24th of November I detected at 03:00 hrs., and watched for one hour, a large flashing, pulsating Ea to the south. It was the same Ea that appears on photo 3A, page 20, and was drawn from on January 17th and 18th, 1955. Later, we concentrated our operations upon this southern Ea which hung low over Tucson, to the exclusion of all others.

On the 27th, the U. S. Government resumed its "Atoms for Peace" plan. The U. S. S. R. wanted to know about denaturization of nuclear material. A great science conference was being planned for Geneva for the coming summer.

We could not rid ourselves of the feeling that we were somehow connected with these "Atoms for Peace" proceedings. We were forced later on to see the connection the hard way. We were to realize that our work was in the center of attention in high social circles; it was declared *"top secret"* at the same time. About the same time Einstein was reported to have said it would have been better if he had become a plumber rather than a scientist. We wondered whether he had finally found out about the deceit perpetrated upon him by a Stalinite mastermind in 1940. His age was dying; so much we knew from the practical aspects of Oranur. But the memory of my meeting with Einstein of January 1941 was to be retained with great pleasure.

It had rained far and wide all around us; but it refused to rain at Tucson. We could not quite understand why. On November 16th it had rained in Los Angeles, Nevada and Utah. At three p.m., we learned that it had rained in California into the Mojave Desert. Clouds were again coming into the Tucson basin from the west, but it refused to rain here. The clouds disappeared again or they did not pour out their water. Still, the moisture increased in the soil all around us.

The cooperation of the Air Force was shown in such details as the following: On November 18th early at 04:00 hrs. I was observing the night sky from the deck. An Ea was again seen near the northern slope of the Catalina mountain. I felt clearly being drawn upon. The Ea moved toward the east horizontally then upward. At 08:00 the Catalina mountain range was black with DOR clouds and a heavy DOR blanket covered the northern horizon. At 08:57 hrs. an AAF plane circled the Mt. Catalina region in an elongated ellipse, in accordance with the layout of the DOR cloud. The AAF knew then and let me know that they knew about the *connection between Ea and DOR*. There was no doubt any longer possible regarding the role Ea played in the dissipation of rain clouds. It did not rain at Tucson, though it rained around it, because of a concentration of Ea observing our region.

This fact required serious counter-action.

Establishment of an OR Potential on Mount Lemmon

We felt annoyed at the lack of rain at our base. The idea was conceived to establish a high OR potential on Mt. Lemmon. We would take a cloudbuster up the mountain about 25 miles to the northeast and draw from the Tucson basin toward the mountain. This would possibly help to cause the clouds to pour the rain out.

The whole crew went up Mt. Lemmon on November 18th, the same day the intervention of the physicians was denied in the Portland Court. We made careful observations while driving up as well as down the mountain. There was clearly DOR hanging over the landscape. I had to distinguish the *"Pocket DOR"* from the *"Ceiling DOR."* The first penetrated into the deep ravines and lay heaviest in the bottom of the deep cut-ins. The latter hovered over the landscape with a more or less sharp edge toward outer space. It was dense enough over Tucson to hide the city from view.

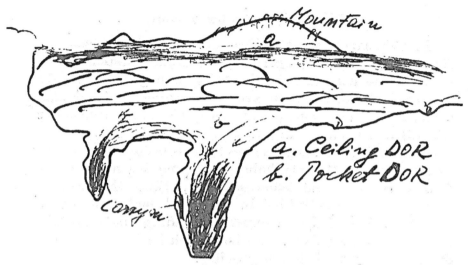

Fig. 24. Ceiling and pocket DOR

It was not difficult at all to remove the heavy "Ceiling DOR." We drew directly from the ceiling; at times we crisscrossed the ceiling, cut it apart, as it were, thinned it out at various places until it gave way. It disappeared after a draw of about 20 minutes.

It was at the moment more important to probe the atmosphere than to accomplish technological rain-making feats. Still, we felt angered when it did not rain. We learned that Ceiling DOR looked *black* seen from above and sideways; that it looked *white* seen from below when we passed through it.

Pocket DOR eats out the rocks, changes them into a brownish powdery mass, clay brownite. Natural erosion by running water on the other hand causes clean, smooth rounded surfaces on the rock.

While we were on Mt. Lemmon a jet flew low above our cars. Also, there were twin motored research planes around following our drawing operations.

On November 21 a huge Red Fireball crashed in Cruxton, Minnesota, according to a Hollywood radio report on KTUX.

"The Right to be Wrong"

In February 1954, the *New York Herald Tribune* carried an article by Samuel Hopkins Adams defending the right of the natural scientist to be wrong without being molested. I had not been wrong regarding the basic characteristics of the energy in the organism and atmosphere which I had begun to discover in 1935. But I had always maintained my privilege not to fear to make mistakes, not to fall into the Babbity habit of not moving along out of fear to displease the fellow scientist. Since the inception of my teaching activities in Vienna I had stressed the point: One must be an expert in finding one's mistakes and in correcting them. Without such basic rules of conduct, basic research is not possible.

The following incident will prove the point. It will, so I hope, impress upon the reader the meaning of what is truly entailed in the term "open-mindedness."

On the 28th of November, 1954, all participants of the expedition made an observation which has remained unsolved to this very day. There were two equally possible interpretations of the factual observation; one was applicable to the classical, the other to the new, so very problematic point of view which maintained that spaceships were visiting our planet. The point to make here is this:

Classical knowledge may all be wrong, such as with the perfect Copernican circles, the ellipses of Kepler, the empty space of Einstein, the airgerms of the Pasteurian bacteriologists, the atomic nature of the Universe, etc. To see new things *from scratch*, to expect the impossible to be true, belongs to the emotional equipment of the true pioneering scientist. Whoever does not possess these qualities should not do basic research. Obviously, as with other serious matters, it is not designed for tyros or routine technicians.

On November 28th, before dawn, I was observing the eastern sky with my 3½ inch refractor telescope. Venus stood already as a sickle high in the sky. To the north of Venus I noticed a small star. Viewed with 60x magnification it presented the following shape and structure: It had two dark symmetrically located black points looking like portholes of a ship. The object had the clear-cut cigar shape of a spaceship. I measured its angular velocity roughly by timing its passage through the diameter of my telescope lens. It passed my field of vision from low left to high right in 149 seconds. The movement was synchronous with that of Venus. Its position with respect to Venus that morning was the following, sketched from the telescopic view.

Position of UFO Nov. 28 and 29 with respect to Venus, 6:30 am, observed with 3.5" Telescope.

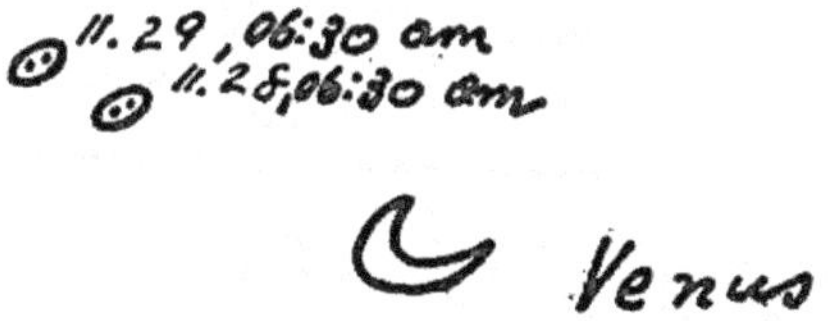

○ 11.29, 06:30 am
○ 11.28, 06:30 am

Venus

Fig. 25. Position of UFO with respect to Venus, November 28th and 29th, 1954

I had no instrument at hand to measure the dislocation from the previous day, but there was no doubt that the cigar-shaped object had moved the following morning, November 29th, 6:30 a.m., somewhat further away from Venus against the line of the ecliptic, toward the celestial Northpole. The atmospheric background count at that time was somewhat erratic, between 50 and 100 CPM.

The object looked like a cigar-shaped spaceship. I refused to accept the notion, but its being farther away from

Venus in a direction opposite to the ecliptic path stuck in my mind. It was at this moment that I had to remind myself of my own principles of conduct in basic research: To hold on to an observation without discarding it, but also without falling for it into uncorrected error; *to keep the matter pending;* to assume that all knowledge could be wrong, or that the Spaceship Theory was wrong. To keep things pending is important.

Fig. 26

I made a few sketches of the obscure thing in my log book and let the matter rest. McCullough who heard about it and saw it after his return from an inspection trip north thought it was or could be Saturn; but he was not certain, and there was no astronomical chart at hand to find out.

We felt it unwise to call at the University for information. The hostility there was already obvious, at least as far as the weather work was concerned. Also, I had learned to keep my new problems away from routine science.

The following picture was taken at a special occasion back in 1946. It demonstrates the helplessness of science and the need for freedom to make bona fide errors, i.e., the "right to be wrong." I leave the photograph unexplained in order to prove the right to be wrong without being molested.

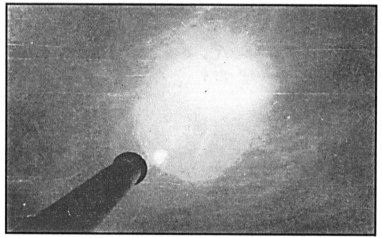

Fig. 27. What does this photograph depict?

That same evening the sky to the west was covered with heavy rain clouds in the form of a high cloud bank. On November 30 the cigar-shaped object could not be observed since the sky was cloudy. The clouds were drifting toward the northeast. The atmosphere was pleasant, like ocean air. Moisture in the atmosphere had climbed again to 67% R.H.; the CPM were high, 80 to 100. The Oracle region, 40 miles away, had greened further; there were also jets overhead. We had the Cloudbuster, located in Yuma in spring, transferred to Tucson. We were now *drawing continuously with two cloudbusters from the south-*

west. We had completely overcome our fear to draw too long. One could not accomplish in the desert in days of drawing what over green lands was possible to achieve with an hour's drawing operation. The observations of Ea and of cloud formations were now reliably integrated. I shall return to this integration soon enough. But first let us return to our mysterious "spaceship" near Venus. It moved according to our observations between December 1 and December 17, 1954, in the following manner with regard to Venus.

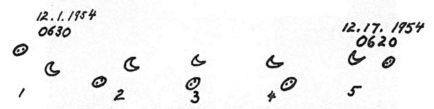

Fig. 28. Position of cigar shape relative to Venus between December 1st and 17th, 1954

It is only through the integration of various extraordinary events that our problem deserves attention. The object swung around Venus and overtook it until on December 17th it was located at 06:20 hrs. to the south of it. Venus had moved westward in a straight line along the ecliptic. Was Saturn swinging around Venus? Or had a spaceship taken off Venus before November 28th toward the north, first removing itself from the planet, then out of its field of gravity and finally swinging southward?

In face of a rigid, doctrinaire, self-appointed, ready-to-kill hierarchy of scientific censorship it appears foolish to publish such thoughts. Anyone malignant enough could do anything with them. Still the right to be wrong has to be maintained. We should not fear to enter a forest because there are wildcats around in the trees. We should not yield our right to well-controlled speculation. It is certain questions entailed in such speculation which the administrators of established knowledge fear: Is it possible, thinkable, that some of the so-called planets are not planets at all: That they are something entirely different? Something unheard of? We do not know and cannot tell. But in entering the cosmic age we should certainly insist on the right to ask new, even silly questions without being molested.

On December 6th, 1954, we experienced two shocking incidents which seemed to bear out the truth and appropriateness of attitudes in basic research such as were mentioned above.

The Breakdown of a Spacegun Operator

December 6th, between 09:35 and 11:00 hours one operator was on order to dedorize the zenith region which was very black. A jet plane drew a vapor trail around the zenith in a wide circle where it was dissolving quickly. In some places the vapor trail did not develop at all. This pointed to something serious going on there.

At about 10:30 hrs. another operator, who had an excellent sense of perception, was moving her hand with fingers outstretched skyward up and down; while she was doing so, she caught a "field" of energy; this was known to us from many experiences during drawing operations, coming apparently from the region of the ecliptic, from a place where Venus must have approximately been located at that hour. A Geiger Counter was brought up and corroborated the subjective sensation of DOR streaming down by showing a count of 400 CPM which rose to 600, 700 and

800: *An Ea was doubtlessly in the sky high up in the region above.* The two other operators were told to draw with the second cloudbuster from the same region. One of them reported "DOR is coming down strong" with a very bitter taste, much stronger than usual and *tasting like offal.* The GM gave 700 CPM. We apparently got the exhausts or whatever else it may have been right down on us. The experience was with us again on December 7th and 8th. On the 7th, the same operator was drawing from the Ea again at 11 hrs. He complained about DOR coming strong and "sour." He felt a crippling sensation in his right leg. (The radio gave at the same time a continuous static noise on all bands.) His right side seemed to go into paralysis. The motility was impaired. However, he recovered soon. The pupillary and other motoric reflexes were unimpeded; so was the sensitivity to tactile sensation.

However, the following day, December 8th, 1954, at 08:30 hrs., while he drew DOR, the paralysis struck again. He came into the quarters sick, purple in his face, hardly able to move his right leg, limping, and with little motility in right arm and hand. A warm soaking bath was taken right away. Warm tea and Southern Comfort liqueur removed the paralysis to a great extent. However, I knew it was serious. The operator had developed a progressive paralytic anorgonia of the right side, under the stream of DOR, still in a functional state, but certainly in the direction of a possibly lasting structural paralysis and nerve atrophy. Handshake was weak on the right side; the right corner of the mouth was drooping and immobile, with right leg immobilized. We advised the operator not to go to any neurological hospital, since we knew well the practices of so many in this profession who after decades of painstaking medical research still were ignorant and determined to remain ignorant of the functional, anorgonotic phases preceding full-fledged neurological disease. It would have been irresponsible to deliver an important worker in our field to such quackery. He went on vacation back to his family, recuperated slowly and returned to work still somewhat impeded late in January 1955. We were happy to have him with us again; however, he was not permitted to draw DOR again. We also thought that in case any neurotic administrator of official quackery in medicine should again declare *"OR energy does not exist,"* we certainly would publicly recommend detention in an observation station for some time of that kind of human animal. Somewhere, some time an end should be put to the nuisance activities on the social scene of frightened biopathic individuals, no matter in what profession or position.

Here is a report on the medical examination after his return to the base on January 28, 1955, by Eva Reich, M. D.:

> Generally much improved over status at onset of neurological difficulty on 12-8-1954.
>
> No new symptoms have arisen. However since return to Tucson on 1-18-1955 has not progressed as steadily, feels this is "at standstill." *Thinks he reacts to Orur operations by dizziness, and return in foot weakness after about 1 hour. Expressed fear of cloudbusting effect on himself.* Still clumsy with right hand, for instance drops a cigarette if he relaxes while holding it. Still stumbles over words, feels mental clarity not impaired. Describes episodes of "waves moving upward" over occiput occurring in past 6 weeks. Tires when walking, can't run well because of right leg weakness. Knows eyes are his main area of difficulty now. Crying is still difficult.
>
> BP 110/60, Pulse 80.
>
> Color—pink, not sallow in contrast to before attack.
>
> Able to wrinkle forehead. Pupils quite constricted, react promptly to light, but are *definitely veiled*. No longer has nystagmus or dizzy spell on moving eyes in all directions. Accommodates all right. Assumes a staring expression when moving eyes (rest of face

becomes masklike). No residuum of facial nerve motor weakness seen. Still is a bit *hypersensitive to pin prick* on right half face.

Lungs are clear except for some mucous rhonchi in right upper lobe. States he may be getting a cold. Has slight runny nose, reddened pharynx. Wax in right ear, left shows old site of perforation. Handgrip slightly weaker on right. *Triceps reflex, right, still hyperactive,* but biceps reflex is equal. No longer has positive Hoffman reflex. Has slight difficulty in Finger-to-Nose test to left. Negative Romberg. Definite weakness of right plantar flexor muscles, some weakness in calf and thigh. Has definite ankle, but not patellar clonus on right. Right sided deep reflexes of leg are hyperactive over left side (knee, ankle).

RMC states that in Army the reflexes in legs were *hypoactive. Positive Babinski response on right side.* Hypersensitive to pin prick on right. Has normal proprioception.

Impression: Healing of nerve centers which had weakened biologically during DOR attack.

No signs of anorgonia now, two sides of body equal in warmth and field.

Residual neurological sequelae of right-sided hemiplegia, improving.

?? Old perforation of left eardrum. Any relationship to present illness?

Advised to keep out of DOR, and if possible use alternation of baths and accumulator.

An operator at a Spacegun had been paralyzed by DOR while drawing from an Ea. This fact was established beyond doubt.

The Arrival of ORUR in Tucson, December 14, 1954

On December 7th it became clear that drought clouds amidst rain clouds indicated the presence of Ea. It refused to rain at Tucson. Therefore, that same day the decision was made to strengthen our position by getting two Orur-Ra needles down to Tucson. Arrangements were made with the Tucson Hudgin Air Service Co. Dr. Michael Silvert in New York, a former Captain and flight surgeon of the Army Air Force, was designated to direct the transport.

DOR was heavily pouring down on us during the followings days. The Air Force was very active in our territory with research planes and jets. On December 8th we saw six jet planes busy all day. At 14:30 hrs. that day we depressed one spacegun and stopped drawing altogether at 16:45 the following day. December 10th was a beautiful DOR-free morning. It had rained the night before in the southern Tucson region for two hours with 0.33 inches of gentle Oranur rain. I was worried the rain may have drowned the proto-vegetation. Rain, good for already germinating vegetation, meant death by drowning to Orene. *Moisture at a distance was the necessary condition for the primal, proto-vegetation.* The moisture had gone down to 9% R.H. on the 9th, but had climbed again to 45% on the 10th. It rained beginning 17:40 hrs. on the 10th over the southern Arizona Desert. Heavy blue-gray clouds covered all of the sky.

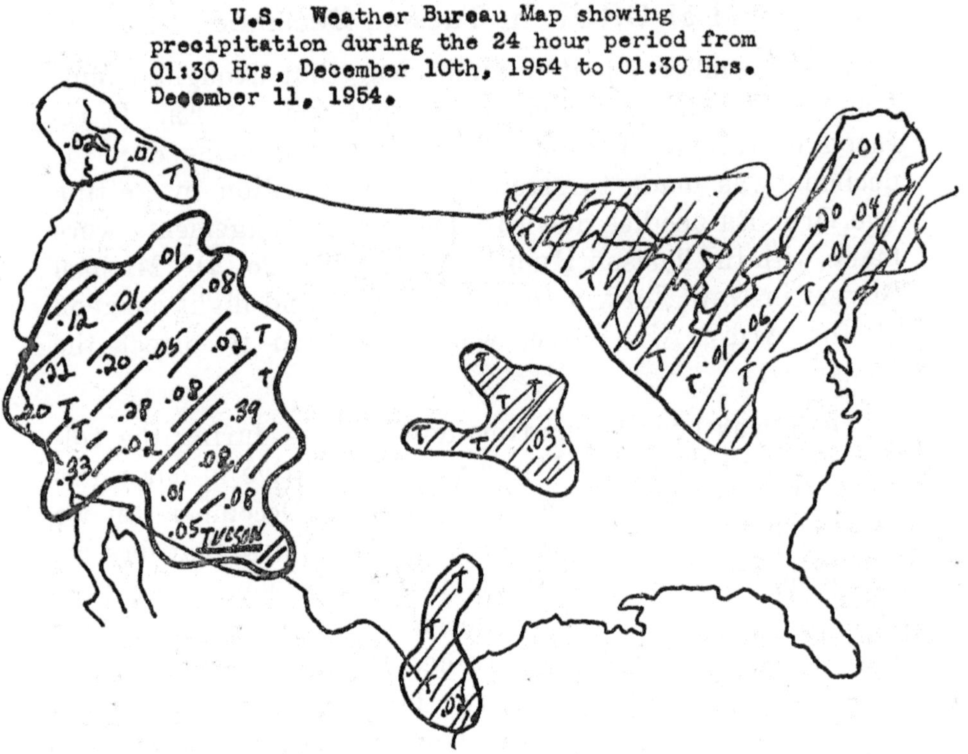

U.S. Weather Bureau Map showing precipitation during the 24 hour period from 01:30 Hrs, December 10th, 1954 to 01:30 Hrs. December 11, 1954.

Fig. 29

We took a rest over the weekend in preparation for things to come. On the 11th of December the atmospheric moisture held its full 80% R.H. in the morning. The sun shone all day, brilliantly. All U.S.A. was clearing that day. The relative humidity went up to 90%, 08:00 hrs., came down to 32% at 14:00. On December 13 the R.H. was only 50% in the morning at 08:00 hrs. We had reinstated the cloudbusters to continue drawing of moisture on December 12, 18:00 hrs. They were depressed again on December 13, 08:00 hrs. to keep the sky clear for the landing of the plane carrying the Orur material.

It was an indication that the atmospheric conditions improved, when after a heavy DOR period, the clouds kept together in a self-regulatory manner without cloudbuster operation. When clouds dissipated again in drought fashion, we knew that the situation had worsened, that it turned downward again toward desert development. Therefore, it was logical to assume that some influence was exerted upon the atmospheric energy when the self-regulatory, cloud-forming power declined.

I was inclined to assume that Ea was the concrete counter agent, counteracting our DOR removal. Ea was probably not capable of standing up in a DOR-free, sparkling, fully-functioning OR atmosphere. It could, so I assumed, exist only where the atmosphere was dorish. From this it followed that if the Ea were manned by intelligent living beings, they themselves must somehow not be capable of holding their machines in operation in a DOR-free atmosphere. This idea was supported by the fact that Ea were often absent on nights after a successful draw followed by brilliant, summery blue OR days. This idea may be wrong, of course, but as a possibility it should be mentioned. The fluctuation of the relative humidity and OR-DOR balance, of heavy rain clouds as against fuzzy drought clouds was sharp and striking in those days before the arrival of the Orur. I could not rid myself of the impression that it worked like a tug of war between two rope-pulling parties.

Dr. Silvert had been instructed to have Tom Ross bring the Orur material down by truck to the airport at Lewiston, Maine. The truck had to be adjusted with a long 2 x 4 to keep the Orur effect as far as possible away from the driver's seat. The Hudgin Air Service was to fly up from Tucson to Lewiston and to install the material. A special container of wood, looking somewhat like a football was constructed, hollow inside, with a suitable opening to insert the Orur material. It was attached to a strong nylon rope 100 feet long. The other end of the rope was arranged in such a manner that the container could be kept trailing in flight at a distance of 100 feet.

One operator was standing by at the phone at all times during the flight. These precautions were very necessary. It was the first time that such a towing was done of active material which did not tolerate metal in its vicinity. We did not know how the flight itself, i.e., the friction with the bypassing air would affect the Orur. It did affect it; this was shown upon arrival.

We did not know how the atmosphere would react to the transport of Orur. It did react to it in an interesting, but little understood manner.

What worried me most was whether the Orur material would directly or indirectly affect the navigating instruments. They did affect them.

Dr. Silvert made the mistake of carrying the lead container with him in the cabin. The following instructions were sent to Dr. Silvert on December 8th:

Instructions for Towing of Material (ORUR) from Maine to Tucson, Arizona, Given 12-8-1954

1. *Never take the material being towed into airplane.*
2. Keep the material being towed away from metal; distance of at least 5 feet, further whenever possible.
3. Keep persons away from material.
4. *Be observant of personnel* in plane during flight, pilot, etc., for biological reactions, possible excessive redness, etc.
5. In case anything goes wrong, land and telephone Orgone Institute, Tucson, Arizona, Tel. No. 3-8263.
6. Place material in safe place away from metal with warning danger signs.

7. *In case towing container is lost*:
   Drop smoke flare, locate the area.
   Land nearest place, inform State Police.
   Telephone Orgone Institute, Tucson, Arizona, or Hudgin Air Service, Tucson, Arizona, 3-1121.
8. Keep watch on tow container during flight.
9. Keep observation on instruments during flight for possible reaction.
10. *Observe atmosphere during flight* when material is being towed from Maine to Arizona. It is unknown what effect the material, being towed rapidly through the air may have on atmospheric conditions.
11. *Landing and Taking Off*: Have plastic towing tube extend 4 or 5 ft. back of plane. Pull towing container back away from plane as soon as possible after landing. Let out away from plane as soon as possible after taking off.
12. *Guard material*: When plane on ground, have material, away from plane, away from metal, guarded with Danger-Keep-Off signs posted. (If signs are metal have them 10 ft. away from material.)
13. *Landing in Tucson*: Pilot should inform Hudgin of arrival time: with *as much advance notice as possible*. Hudgin *inform Orgone Institute of landing time*. Landing will be at Gilpin Airport, Tucson.

The first telephone report came in from Pittsburgh on December 12th, 19:00 hrs.:

December 12, 1954. *Dr. Silvert called at 19:00*—from the Greater Pittsburgh Airport, Pittsburgh, Pa. Towing gear riding smoothly, no trouble * * * No biological reactions on pilot or men in plane. CPM in plane around 20, go up to 50 or 60 when towing container brought near plane when landing or taking off * * * *A deep blue in sky seems to follow plane in flight with spotty cloudiness* * * * *weather predictions are changing fast.* S. is spending night in Pittsburgh * * * weather may not permit take-off until tomorrow afternoon * * * do not now expect to arrive in Tucson until Tuesday. Airfield manager at first reluctant and uncooperative in regards to idea of necessity for the guarding of material. *Then became extremely cooperative with State Police volunteering to guard material* * * * *Material safe on airfield with signs,* etc. * * * Will let us know ahead in plenty of time regarding their *arrival in Tucson.*

This telephone report relieved greatly the worry about possible plane disaster.

I had instructed to land immediately at the nearest airport should the pilot show signs of purple discoloration of the face, bloodshot eyes, dizziness or high pressure in brain. I had intended to examine the pilots but desisted upon consideration of the fact that such request was inappropriate. A few questions had calmed my worry. The men seemed basically healthy. I had instructed the pilot personally at the airport before take-off of the strict rules to follow during the transport.

The arrangements had been made in a careful manner, avoiding unnecessary publicity and unpleasant unanswerable questions such as "What does the Atomic Energy Commission think about Orur?" We could not possibly tell them that it knew nothing about it or would even be in the way with all kinds of rigmarole. Or whether "The Weather Bureau had any opinion about it"; it had not at all, but it would have been most unpleasant to tell a news reporter without being written up as a crackpot in some newspaper. Some publicity burst into the open only in

January a few weeks after the completed transportation. My priority was thus secured.

Report: Transfer of ORUR to Tucson, Arizona, December 12-14, 1954

1. *Dec. 12, 1954*—0500 to 0800: Orgonon, Rangeley, Maine, to Lewiston, Maine—Trucked by Tom Ross at ends of two 2 x 4s connected by a crosspiece, projecting at rear of truck. Lead container (LC) carried at one end, OR at the other. Ross complained only of "some blocking in back of neck", and looked well. CPM in truck cab after removal of OR and LC (lead container): 36. No opportunity to test before removal.

2. *12-12*—0800 to 0900: Transfer of OR and LC to plane at Lewiston—On approaching the OR to within 3 or 4 feet, CPM rose rapidly to over 60,000. CPM of LC, 40. CPM of double plastic container 36. LC and double plastic container placed in plane's baggage compartment just behind and beneath rear seat. The two OR units in individual plastic containers packed in cotton in wooden egg; this placed into canvas bag and laced securely (later also taped when lace frayed when dragged behind plane while taxiing—only dragged this first time, thereafter carried out and back by Silvert while reeled by Bolman, second pilot). The canvas bag was towed at end of 100+ foot nylon cord (1000 lb. test) leading from plane through mycarda tubing projecting 4 feet from tail tip. The cord extended forward through rubber garden hose into cabin, then on an improvised reel hand-held. In flight, the tow exerted a strong pull, and required considerable effort to reel in. Silvert pulled in the cord while Bolman reeled up the slackened cord.

Pilots: Henry Hudgin, Frank Bolman.

3. *12-12*—0900 to 1210 (EST): Lewiston, Me. to Harrisburg, Pa.—OR reeled in for take-off then out full as soon as plane airborne. At 10 to 15 min. intervals, the following checks were made: location (altitude, course, place if known), tow (checked by tugging), weather, CPM, instruments and people. (Detailed record in tabulated form available.) CPM ranged between 16 and 26. Weather, at first strongly sunny and "clear", gradually became blue-grey cloudy. When the pilots were advised to look for and report any unusual instrumental or individual reactions, Hudgin smiled, "I'm a bit that way already", circling his finger at his temple. A blue haze soon appeared all around. Silvert became sleepy at 1005. At 1100 Hudgin reported burning of eyes, and "a pain and ache in my stomach", which he attributed to hunger, having had no breakfast. Altitude 3000-7000 ft. depending on ground visibility, due to lack of radio (plane new, radio not yet installed). Flight by contact, with no flight plan. Instruments and people showed no other effects.

4. *12-12*—1210 to 1325 (EST): Stop at Harrisburg for gas and food—As OR reeled in, both for landing and take-off, CPM began to rise when about ⅔ in, reaching 60 with OR at end of mycarda tubing, about 5 ft. from tail tip. Unreeling, CPM would fall again to about 20. This recurred regularly, except that later, the CPM would not fall so rapidly or completely, as will be noted.

5. *12-12*—1325 to 1600 (EST): Harrisburg to Pittsburgh, Pa.—Weather closing in. Hudgin unable to locate Allegheny County Airport in Pittsburgh, flew on and made an emergency landing at Weirton Airport, then flew on to Wheeling and into fog, turned back and landed at the Greater Pittsburgh Airport. In my opinion, Oranur-induced confusion led to difficulty in locating the first airport, then poor judgment in flying further. The pilots later agreed that this had happened.

6. *12-12*—1600 to *12-13*—0955 (EST): Stop at Pittsburgh for weather and overnight. After unreeling OR after landing, CPM fell to 20, but more slowly than before. That evening, Hudgin was purplish-red in the face. Bolman was anxious and talkative, Silvert was restless and hot all over. Weather reports indicated that we might not be able to leave until the 14th, or at the earliest, late on the 13th. However, weather changed unexpectedly, so that we departed early on the 13th. Airport officials were at first critical at our landing, because of the general objection of airlines to private planes. Then they became interested in the OR precautions and were very cooperative with supplying warning signs, and offering an all-night state trooper guard which was deemed unnecessary. There was a small amount of snow at this field. Silvert insisted that all take long showers before going to bed. I consider that the unexpected weather change was an Oranur effect in the sensitive atmosphere in the East.

7. *12-13*—1005 to 1230 (EST): Pittsburgh, Pa. to Danville, Ill.—The Pittsburgh control tower held the plane up for about 10 minutes while airliners were landing or taking off. The OR was reeled in all this time, the longest period of the entire trip. Shortly after take-off, Hudgin reported that the compass was deviating 20° and had thrown him considerably off course before he caught it. This gradually subsided and after an hour, the compass read true. Hudgin also noted that when Silvert leaned forward, the compass deviated significantly; when Silvert leaned back, the compass returned. A thick blue haze appeared around the plane, and this continued the rest of the way as long as daylight permitted observation. Flew generally at 5000 ft., under a high overcast. The compass deviation appeared to be an Oranur effect induced by the over-long nearness of the OR to the plane during delay in take-off. The observation of effect of leaning forward on the compass suggests that the Oranur effect on the compass

may have been mediated through the organisms of the people in the plane, at least in part, rather than directly. Thus, the change in fields of living organisms in turn affected the field around the compass. The thick blue haze was unusual; the pilots said it was more suggestive of Los Angeles, so that this is another probable Oranur effect.

8. *12-13*—1230 to 1300 (EST): Stop at Danville, Ill. for gas—Fairly cold, no snow.

9. *12-13*—1300 to 1520 (EST): Danville, Ill. to Kansas City—Pilots and Silvert felt a bit euphoric. "We know where we are now." The thick blue haze increased in intensity. Increasing cloudiness caused a change in altitude from 6000 ft. to 2000, leading to some irritation because of lower speed and increased roughness. (Plane speed with reference to the ground is greater at higher altitudes, and planes fly smoother above clouds than below.)

10. *12-13*—1320 to 1355 (*MST*): Stop at Kansas City for gas. Considerable "smog".

11. *12-13*—1355 to 1630 (MST): Kansas City to Gage, Okla.—Hudgin rested in back seat while Silvert took pilot's seat. Crossed Mississippi River at Hannibal, and left troublesome clouds behind at about the same time. At 1500 saw two vertical rainbow arcs up ahead, about 45 min. after leaving the Mississippi R. and while flying 8000 ft.

12. *12-13*—1630 to 1650: Stop at Gage, Okla. for gas. Cool. Some impatience and irritation. Bolman complained to Silvert about Hudgin's "stubbornness".

13. *12-13*—1650 to 1800: Gage, Okla. to Amarillo, Texas—We all observed that the temperature above the ground was consistently higher than on the ground. Hudgin said that while this is often seen, it was unusual

for it to last so long. Thus, another Oranur effect is a higher temperature around the plane. Flying about 6000 feet. Beautiful sunset.

14. *12-13*—1800 to 1830: Stop at Amarillo for gas; might not find gas at Roswell, N. M.

15. *12-13*—1830 to 2130 (MST): Amarillo to El Paso, Texas—Flying 12,000 ft. Abdomen feels swollen, uncomfortable. We had eaten little all day. CPM now a little higher, up to 40, as against a former 20. High scattered clouds, probably 20,000 ft. or higher.

16. *12-13*—2130 to 2300 (MST): Stop at El Paso for gas, food. Phoned OIRL at Tucson. (Wired at Lewiston, phoned at Pittsburgh, wired at Kansas City.)

17. *12-13*—2300 to *12-14*—0100 (MST): El Paso to Tucson, Arizona—Airway beacons were easier to see at night than ground landmarks in daytime. Also pilots' familiarity with region made a noticeable difference in their attitude, so that they were much more at ease. Silvert felt aches in back of neck and right wrist, some malaise and unrest, then felt better. Pilots feeling fairly well. Flying 11,000 ft. Light high cloudiness. Instruments probably working well, although pilots were following beacons and familiar areas. Good ground visibility.

18. *12-14*—0100 to 0825 (MST): Tucson Municipal Airport. Met Mr. Moise. Stop for overnight. Turned over LC to OIRL, and removed from plane by Moise.

19. *12-14*—0825 to 0840 (MST): Tucson Municipal Airport to Airport in north end of Tucson. Hudgin and brother in plane with Silvert.

20. *12-14*—0840: Landed north end of Tucson. OR turned over to WR.

Summary:

1. ERROR: Silvert took the lead container along in plane, risking loss of OR, personnel and plane by intense Oranur interaction between OR and lead, with effects on people, instruments and ignition system. For short periods, LC and OR were within 25 ft., but mostly were 100+ ft. apart.

2. Effects on instruments: 20° compass deviation after leaving Pittsburgh after delay in take-off. Deviation gradually subsided in an hour. Deviation when Silvert leaned forward suggests that compass effect may have been at least partially mediated by field changes in living organisms.

3. Effects on people: Some confusion while approaching Pittsburgh. At various other times, drowsiness, irritability, impatience, burning of eyes, aches and pains, euphoria.

4. Effects on atmosphere: Weather cleared at Pittsburgh unexpectedly, permitting an earlier departure than anticipated by perhaps 24 hours. Thick blue haze around plane, especially marked out of Pittsburgh. (Could the marked industrial DOR concentration in this city have interacted with the OR and LC?) Higher temperature around plane in flight than on the ground, lasting longer than expected.

5. CPM: In plane with OR reeled out, on ground or in the air, about 20, rising to 40 near end of trip. As OR reeled in, on ground or in the air, CPM would rise to 60, beginning to rise when OR ⅔ reeled in. Approaching OR on the ground, CPM would begin to rise within 25 ft. to over 60,000 within 3 or 4 ft. Toward end of trip, CPM would fall more slowly than before, when OR unreeled.

6. Effects on airport personnel and others: Tom Ross reported "some blocking in back of neck" from his long truck trip from Orgonon to Lewiston. There was a general interest with seriousness, and willingness to cooperate in safety precautions. Several remarked, "I never saw a rig like that before". There was no apparent EP, although the manager appeared a bit over-zealous in bringing in 3 state troopers to see about setting an overnight guard, although we had not requested one. (Perhaps he was right and we were wrong!)

7. Finally, Pilots Hudgin and Bolman showed much ingenuity and serious interest in installing the towing mechanism at Lock Haven, Pa. The tow showed itself to be fully adequate in operation, as originally conceived by WR. The experience was a rich one for all concerned.

                Respectfully submitted,

    /s/ Michael Silvert
    ORGONE INSTITUTE, Oranur Weather Control
    Michael Silvert, M.D., Operator (Medical), N.Y.

Excerpts from News Report—January 10, 1955:

"Tucson newsmen are buzzing with curiosity over recent flight made by Al Hudgin * * * operator of a flying school here.

"Hudgin flew back to Maine and returned with a small amount of radioactive material * * * so hot it had to be dragged behind his plane on a fifty foot long wire.

"Landing at the greater Pittsburgh airport, whatever it was Hudgin was dragging queered the operation of the air force radar screens there. Airport attendants were warned to stay fifty feet away from the radioactive package.

"Recipient of the hot material in Tucson is Dr. Wilhelm Reich * * * a renowned Viennese phychiatrist. He is the author of several noted books on psychiatry and is currently conducting experiments of an unexplained nature at the Orgone Research Institute Laboratories on McGee Road.

Chris Cole, reporting"

Televised February 18, 1955, KEAN, Tucson, Arizona:

"Some weeks ago, I told you of a bundle of aggravated uranium * * * flown from Maine to Tucson by private plane on the end of a long wire * * * because it was believed the stuff was too hot for normal transportation agencies to handle.

"The bundle was delivered to Doctor Wilhelm Reich, a noted Viennese scientist who is currently conducting a most unusual experiment on a ranch north of Tucson.

"I visited Doctor Reich and discovered he has plans for our city and the desert surrounding it. The discoverer of a natural force he has termed Orgone Energy, Doctor Reich is attempting to battle Arizona's drought with his own methods. He is hoping to make rain.

"A graduate of the University of Vienna, Reich is the founder of the Orgone Energy Laboratories in Rangeley, Maine. For seven years he worked as first clinical assistant to Sigmund Freud in Vienna and is the author of several books dealing with psychiatry. From 1934 to 1939, he lectured at the Psychological Institute of the University of Oslo, Norway.

"Declaring he believes all deserts to have been man made during the past twenty five thousand years, Doctor Reich told me he is using two machines of the type you see here to create clouds and bathe Tucson with gentle moisture.

"The gadgets look something like anti-aircraft guns and cost approximately eight thousand dollars to build. I failed to discover what part the imported uranium played in the action of Doctor Reich's cloudbusters • * •. But did learn that each of the machines has a long cable extending into a deep well on the ranch property.

"After satisfying himself about the possibility of rain making here, Doctor Reich assured me, he will head for Los Angeles to see what effect his cloudbusters will have on smog conditions along the west coast."

ORUR-Highly Excited

The plane had landed at Tucson Municipal Airport at 01:00 hrs., on December 14th, 1954. The Orur material was stationed there until the morning. Vle saw the plane finally landing at an auxiliary airport near our base with the container trailing neatly behind. We saw the container being pulled in before landing at 9:40 a.m. on December 14th, 1954. We measured the activity right away after arrival at Little Orgonon.

Activity within the container

    at 100 feet .... 100,000 CPM
    at 60 feet .... 100,000 CPM

OR outside container

    in ground .... 100,000 CPM
    in 1" lead .... 100,000 CPM

Container alone *before* use .... 750 CPM
                 after use .... 500 down to 200 CPM

Cotton from wrapping .... 400 CPM steady.

The Orur material had calmed down on December 15th to an activity of

200 cpm at 5 feet	(100,000 on the 14th)
1200 cpm at 3 feet	(" " " ")
2400 cpm at 1 foot	(" " " ")
80,000 cpm on Contact	(" " " ")

The Orur material was highly excited; at Orgonon it had yielded only low counts if not in lead container. Orur calmed down slowly during the following 24 hours. Its field shrank gradually from 100,000 cpm at 100 feet to 200 cpm at 5 feet at 16:00 hrs. But close by it still gave without metal 100,000 cpm. It appeared that the transport through the OR ocean and the atmosphere at about 250 miles per hour over 3000 mlies had had an exciting effect upon it. It had lost its zero reaction. Its field stretched upon arrival at Tucson to several hundred feet. It behaved, as should be expected of life energy, like an excited animal that calms down after a while. After nine hours the field of excitation had shrunk from 300 to 5 feet.

During our operations a few AAF planes were circling overhead, probably testing the background count vertically in the atmosphere.

We had all been so busy with the reception of Orur that we had not given any thought to the main issue, the Ea problem. However, startling events that very same night and in the late afternoon again shocked every one of us into sharp awareness of what we were facing somewhere above us.

The Ea Battle of Tucson, December 14, 1954 — 16:30 Hours

On December 14, about 16:30 hrs., a full scale interplanetary battle came off; a battle which would have appeared incredible as well as incomprehensible to anyone who knew nothing about the Ea problems or who adhered to the illusion that neither Ea nor Cosmic Energy existed. The *"Ea Battle of Tucson"* fitted without difficulty into the pattern of our past experiences from previous encounters with Ea. It demonstrated in a concentrated form, as it were, what had gone on before and what we could expect, roughly, of future encounters with our uninvited visitors from outer space.

At the time of the landing of Orur at Tucson Municipal Airport, two operators present there witnessed the dropping of about 30 flares around the region. Such flares had been seen during the Ea operations at Orgonon. Dr. Silvert, too, reported to have seen many flares (and AAF planes) upon arrival at Tucson. This was the prelude to more serious events on the afternoon of December 14th.

Two staff members returned from Tucson in the late afternoon in very bad shape. One felt sick and appeared dulled, somehow out of balance. The other looked pale, felt severely nauseated and complained that *"the car had been so very bad."* They reported that the man in the hardware shop where they had bought some things had looked pale and had acted confused, as if paralyzed. One operator was of the opinion that we were dealing with a special kind of DOR attack.

At 16:30 hrs. a tremendous black cloud, looking like smoke from a huge fire arose over Tucson. The cloud later spread out and became deep purple with a glowing kind of reddish hue. The Geiger Counter registered at that moment 100,000 CPM. The situation appeared threatening, bizarre and frightening at the same time. It was soon clear that it could not be a fire. Everything pointed to an attack by Ea low over the city of Tucson.

Fifteen minutes later at 16:45, according to the written protocol, about a dozen Air Force planes of various kinds appeared over Little Orgonon. Vapor trails of jets did not hold together at all. We all suffered from nausea, quivering, pain in the upper abdomen and discoordination of movements.

I alarmed all operators and instructed them to draw with one spacegun from zenith and with the other from the cloud over Tucson 8 miles away. The first was to weaken any Ea present at zenith; the other was to remove the dangerous mass of purple, smoky menace from Tucson. The DOR cloud over Tucson began to shrink after a few minutes' operation. But it lasted about 20 minutes until the sky was clear again.

At about 17:18 a single jet plane circled Little Orgonon. Four big B-56 jet bombers gathering from over Tucson flew toward Little Orgonon and passed low overhead and slowly in closed formation at 17:30 hrs. I had the impression that they saluted our base. There was in the situation an emotion of deep concern, determination and gratitude.

Dr. Silvert who was with us on the deck witnessed the total battle and the following salute. My son, Peter, had helped with a megaphone, since my voice had still not recovered from hoarseness, in relaying my orders to the operators at the two spaceguns. Peter also discovered a *silvery disk moving from the region over the city, where the purple clouds had developed, toward the west.* Peter pointed the Ea out to me and I saw it then right away. 1 had forgotten this incident until I found a note in the Log Book referring to it while preparing the material for typing a week ago. Even seasoned operators refused to admit fully what had happened. However, we all knew that something extraordinary and dangerous had happened: *Had the Ea launched the attack upon the municipal airfield in the belief that Orur was still there?* No one can tell.

The GM reaction, too, calmed down with the CPM from 100,000++ to normal again within about two hours. We all took stiff drinks and tried to be as happy as possible under the circumstances. Emotionally speaking, words about the battle with the Ea were slow in coming over our lips.

I cannot tell how many people had died during that hour from heart attack and similar reactions to a sudden DOR attack as severe as the one just experienced.

The Nodes of Ea

By concentrating upon single regions, single objects, single riddles, we avoided confusion and reached efficiency. For example, it was found that an Ea is *energetically* not shaped round but is built like a long stretched-out double cone with two distinct focal points. We shall postpone a discussion of this fact. It has great bearing upon other terrestrial and celestial functions, for example the location of the origin of hurricanes.

Fig. 30. Ea-CM nodes*

---

* CM—Core Machines.

First ORUR Operations in the Desert

On December 16th I organized the first ORUR operation with the spacegun in the desert. We started out with a 10 second operation; we orurized for 100 minutes a few months later at a special occasion on February 15th.

The harvest of new knowledge was rich, over-abundant, thrilling, pregnant with future. The connection between drought and Ea activities had been secured.

The first victories in combating desert and its makers, the Ea, had been won. Deserts were shown to be able to green without rain. *Secondary cholla vegetation like secondary drives in the armor was dying upon return of primordial life. Killer Atom had been converted to peaceful use.*

During the first days of ORUR operations in the fully developed desert, we experienced with great satisfaction the sight of sparkling bluish-gray OR energy enlivening the formerly stale, black, nauseating atmosphere. The way was open ahead of us to proceed toward still richer exploits in the combat of the age-old worst enemy of mankind. The road ahead appeared full of obstacles, it was true, but the balance in our favor seemed secured in the handling of the space problem between the schizophrenic phantast, who mistakes his voices for talks with spacemen, on the one hand, and the frightened little man in every walk of professional life who centers his universe around his chronic constipation and his lack of male or female animal potency. Both types were of the past. The future seemed to harbor a new type of man and woman: The one to whom the omnipresence and activity of a Cosmic Energy, a Life Energy was no problem at all and never had been; who knew the quivering in an embrace of full delight. The man and woman and child who had always known it, and had only wondered at times why they had never heard of it in public lectures, in the movies, in social gatherings, from radio and television, expounded by academic teachers. But they knew well this ever-present, all pervading energy which was life per se gleamed in the eyes of small children, radiated from the bodies of soft-spoken, good-natured black women in the American South as it did in any forgotten village on this earth where the Stalinite killer had not penetrated yet with his evil CCC, conniving, confusing, conspiring to kill life. The return to full sparkling Life of the nauseating DOR clouds after Orurization reminded one of the realities in man's dreams about a revival from death which truly only meant to him being freed from DEADNESS of the emotions. How faithfully he had tried to remove the DOR from bis body by bathing in the rivers of the earth from the Ganges to the Mississippi. Even the water in the depths of the earth's crust seemed to emerge toward the light of day again: The riverbeds in the Tucson region darkened and became moist, as if the moisture from the depths were trying to meet the moisture from the atmosphere over the reviving desert; as if in a brotherly embrace after millennia of separation there was unity in sight again. God seemed to become earthy again.

The January Oranur Rains

Toward the end of December 1954 an Ea attracted my attention by its peculiar movements. On December 23 1954 it moved as follows:

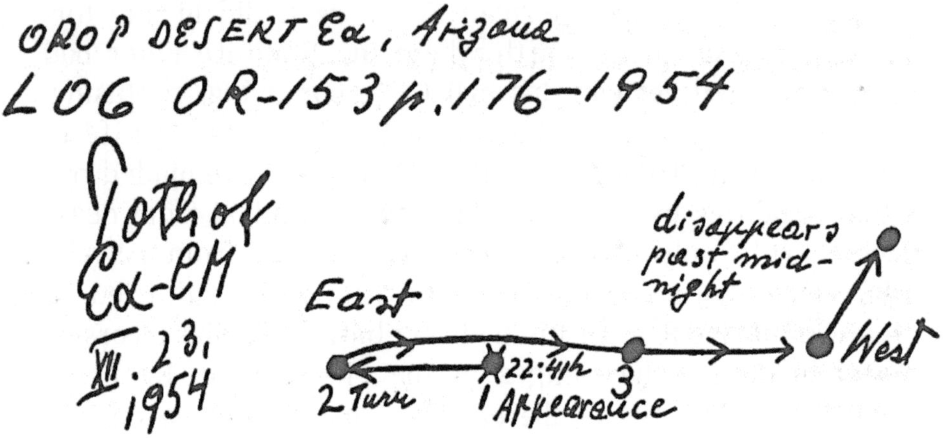

OROP DESERT Ea, Arizona
LOG OR-153 p. 176—1954

Both of Ea-CM
XII 23, 1954

East → 2 Turn ← 22:41h 1 Appearance → West
disappears past midnight

Fig. 31. Ea-CM path

The Ea continued to appear irregularly at various places and at different times of the night. We became gradually used to their presence, sudden appearance, disappearance or absence. We had also learned to react spontaneously to their presence with our organismic reactions which were now better known to us. However, our space-gun operations became rather elaborate as we searched our way further into the unknown territory. Various ways of using Orur developed. One way was to keep the Orur in the lead container stationary near the draw pipes. Another was swinging the Orur to and fro or past the metal pipes. The duration of Orurization varied greatly according to the severity of the DOR emergency, the weather reports we heard regarding local and national developments, etc. We soon learned that the desert atmosphere tolerated rather extended operations, up to 1½ hours, without major upheavals.

The people in Tucson and in the vicinity of Oracle had become aware of the changes that had taken place in the desert. They were much more alert to these changes, naturally, than the people in the east had been, because the changes were much more striking. On December 20 our operations were televised; I had given a brief summary of their functions into the microphone. The film was darkened. It could have been from the presence of Orur in its vicinity. There was also the possibility of foul play. Ten days later, on December 30, a drug agent turned up at Little Orgonon accompanied by a U. S. Marshal who had come "as a friend, not on duty, as an observer only." The drug agent was asked whether he had come on official order. He said no; he wanted, however, verbatim to *"inspect everything."* Thus, he was impertinent enough to expect me to show him my private diary, my poems, my love letters, my microscopy log book, my personal history documents, complete à la Beria. He was, of course, not admitted to our premises and had to leave. It is so very difficult to pin down situations which arise from subterranean evil; however, his visit had some connection with the rumors that went around in town regarding our work.

The filming of our base had stirred up interest. Whether the blackening of our film was due to some unlawful interference, I cannot tell; it was entirely within the realm of the possible. Also, the Boston Circuit Court of Appeals had denied the application by the physicians to intervene with the unlawful injunction. There was no doubt whatever that enemies of our work were busy behind the scenes to find out what we were doing at Tucson. Both, the American chemical industry and the Russian politicos were rather interested. I lived and worked in a glass house, as it were.

January 1955 was the month during which it rained repeatedly and abundantly in the southwestern U. S. A., due to our operations. The transportation of the Orur material of December 14th was reported on Television News on January 10 (seep. 195). A conference with local farmers, business representatives and banking officials was scheduled for January 28th to discuss the problem of how to continue the desert work after our departure which was planned for the end of April.

The Oranur rain on the 3rd, 6th and 7th of January was rich and gently continued all through the night. There were deep water puddles in the streets. The soil was well soaked. There was snow on Mt. Catalina. There were floods in Mexicali. Two thousand families had lived in ever-dry river beds and now had to be evacuated.

The DOR situation became very bad again on January 12. There was a black, heavy DOR blanket at the northern horizon. Three jet planes flew past at 11:00 hrs. from north to south with visible vapor. trails. Suddenly the vapor trails disappeared. There seemed to be trouble in that region, no doubt. We orurized toward that region at 11:15 hrs. for five minutes, stationary alternating with drawing from zenith which was heavy with DOR. At 11:20 hrs. the three jets flew low over our Spacegun as if in agreement. At 17:15 hrs. the zenith was black again. There was no doubt that a new tug of war with the Ea was on.

For the first time one spacegun was installed to draw continually from the region where we knew Ea usually appeared during the night; the purpose was to weaken the potential in *that region in advance.* I't was a bad day. The horizon was black with DOR all around us. Only the sky over Little Orgonon seemed to be clear. The DOR at the horizon did not budge. On January 13, before dawn, an Ea, big, yellow, flashing in the western sky went out in the region toward which the spacegun had been pointed all through the night.

January 13th was, in contradistinction to the 12th, a clear day with high relative humidity, 70% at the base, and according to the radio report 80% at the airport. Vapor trails held beautifully that day. The sun shone brilliantly, not at all desert-like, through a wonderful blue-gray haze. At 13:30 the sky began to cloud over totally with rain in prospect. At 19:15 hrs. the sky was nearly completely clouded over. At 20:00 it began to rain gently; the rain continued all through the night into January 14th. The moisture in the atmosphere had risen to 90% at the base, to 96% at the airport. One of four physicians who had visited us from New York was unable to leave by plane because of the heavy rain. No planes landed that morning. It was reported that such an event had not happened before in years. Flight personnel were amazed.

The Arizona rain went on to Texas and New Mexico with abundant showers. Fog covered the Phoenix valley. A gigantic storm developed in the Atlantic from Puerto Rico to Labrador, as a gale over a stretch of 1600 miles.

I cleared the sky over Tucson at 12 noon. That day clouds were all around the Tucson basin. The planes which were stranded were now able to take off from the Tucson airport.

The two Spaceguns remained centered since 18:00 hrs. upon the Southern Ea which usually appeared around 21:00 p.m. (It was on January 17 and 18 that the two photographs were taken of the southern Ea.)

The following survey conveys an impression of the sightings and operations December 21, 1954 to January 20, 1955. The concentration of our attention on one single Ea bore its fruits:

"THE SOUTHERN Ea—CM"* OVER TUCSON, ARIZONA

(a)

Night Date	Presence Begin (Hours)	Presence End (Hours)	Description	Location, Movement	Cloudbuster Drawing from CM	Result, Remarks
12/21/54	23:00	?	Red—White—Blue Pulsation +++	E—W 10° 15° W of S	—	—
12/22/54	+	+	"	"	—	—
12/23/54	+	+	Yellow Pulsation +++	10° E of S	25 min. 02:05—02:30 1 ClB	Fainter, 02:30 No pulsation; No red flash.
12/24/54	22:41	02:30+	Red—Blue—White Pulsation +++	E—W	+ Orur	Much fainter after draw, fades out, not completely. No pulsation. Yellowish.
12/25/54	23:00→23:25	23:05 02:30+	Red—Blue—White Pulsation +	E—W and up	+ 2 ClB 23:25—24:05	Large blue flare South 23:12. Fainter.
12/26/54	22:45→23:25 24:00	23:25 02:30+	Red—Blue—Yellow Pulsation +	E—W and up	+ Orur 6 min. 23:25—23:31	Fainter, blinks out, troubled, but stays in sky. Humidity rises during OROP to 50%.

*CM — CORE Machine. This log, in rough draft, is at times unclear. Clarification will, if possible, be obtained from original notes at a later date.

Night Date	Presence Begin → End (Hours)	Description	Location, Movement	Cloudbuster Drawing from CM	Result, Remarks (b)
12/27/54	22:30 02:30+	Green-Yellow, steady	15° E of S E-W	+ Orur	CM moves up to E during draw. Fainter, still there at 01:00.
12/28/54	22:20→22:30 22:45 02:00+	Yellow-Red Pulsation ++	Appears at 3° 15° E of S	+ Orur 9 min. 22:56-23:05	Flickers, dims
12/29/54	22:12→23:45 seen with binoculars 23:50 02:00+	Yellow-Blue, little Red Pulsation +	E-W, up	+ 2 ClB 20 min. 22:37-22:57	Much smaller, fainter, no pulsation, no red.
12/30/54	19:05? 21:58	Yellow Pulsation ++	"	+ 1 ClB 20 min. 23:20-23:40	Gets fainter, no pulsation, blinks out more.
12/31/54	Cloudy	—	—	—	—
1/1/55	21:55→22:40 22:50 02:00?	Yellow, blinks, fades out	"	+ ClB 27 min. 22:10-22:37	Blinked out during draw, on return seemed "refreshed," pulsated Red-White-Yellow again.
1/2/55	Cloudy, Rain	—	—	—	—
1/3/55	Cloudy, Rain	—	—	—	—

Night Date	Presence Begin End (Hours)	Description	Location, Movement	Cloudbuster Drawing from CM	Result, (c) Remarks
1/4/55	Clear WR 18:00? to SE? 22:00 02:00	Yellow Pulsation +	–	–	–
1/5/55	21:45 02:00	Yellow, occasional red. Pulsation ++	Appears at 3°, E of S Moves E-W	+ 30 min. 22:00-22:30	Less red flashes, faint after draw; 22:00 bright flare to S, Red-Yellow core, blue glow. Observe "F Point" to West of CM 10°
1/6/55	21:40 Before 02:00	Yellow, occasional red. Pulsation ++	3° E of S Moves E-W, up	1 ClB 21:50-22:20 30 min. Orur + 6 min.-22:27	Fainter if we draw from nodal points 10° to E and W; fades when this point touched.
1/7/55	Rain	–	–	–	–
1/8/55	22:00 Before 02:00 Clouds, clear	Red-Yellow Pulsation ++++	Same	1 ClB 00:20-00:25	It drew from us strongly at midnight. To stop this, we drew from it: very bitter. Fainter afterwards.
1/9/55	21:30 01:30+	Yellow Pulsation ++	–	+ Orur 4 min. 21:50	Fainter sparks out, did not feel bitter points clearly.

Night Date	Presence Begin End (Hours)		Description	Location, Movement	Cloudbuster Drawing from CM	Result, (d) Remarks
1/10/55	21:25	01:30	Yellow Pulsation ++	E to W	ClB to S all night	(Strong DOR in AM)
1/11/55	21:30	01:00+ out 4 times			+ 1 ClB 1 hr. 21:25 23:15	Extinguished 4 times by drawing crossfired at nodal points.
1/12/55	DID NOT APPEAR. Sky clear in South at 21:00. Sky clouded over, rain started at 21:30.					
1/13/55	Rain	—	—	—	—	—
1/14/55	Cloudy	—	—	—	—	—
1/15/55	Cloudy Midnight 01:00		—	—	—	—
1/16/55	Rain	—	—	—	—	—
1/17/55	21:00	01:00	Yellow-Red-Blue Flashes	—	—	—
1/18/55	21:00 21:30	21:15 00:30+	Yellow, Occasional red	—	30 min. 2 ClB "Crossed" in advance	Effective, weakens CM before it appears. Faint, peters out repeatedly.
1/19/55	Cloudy, rain 21:00		—	—	—	—
1/20/55	20:55	01:00+	Whitish	—	45 min. 1 ClB	Fainter, seems farther away.

Compiled by Eva Reich, M.D., 1/21/55.

CHAPTER VIII

BREAKING THE BARRIER

Search for Atmospheric Self-Regulation and the Obstacle in the Way

With the Ea battle of Tucson the first chapter of the orgonomic Ea story reached its end. With the systematic use of Orur in the following operations against DOR, Desert, Drought and Ea the next chapter begins. But before we continue with the report on the remaining four months of the expedition, a few loose odds and ends have to be picked up. A few general points summarizing the first two months should be brought forth, too.

I had mentioned a while ago that it had never been the objective of the expedition *"to make rain over deserts."* The true objective had been to study carefully the conditions which are responsible for drought, desert and starvation of the Living: Briefly, the *"Obstacle in The Way."* It still was *not* the purpose of the expedition to make rain, even after we had broken a drought spell in Arizona of five years' standing and had achieved our first gentle oranur-type rain on December 9 and the following days. The objective was to *find the borderline where our artificial efforts at a new atmospheric technology could end and be replaced by the self-regulatory, self-sustaining laws that govern the behavior of cloud formation, rain cycles, cosmic energy metabolism in the atmosphere,* etc., as they do in the living organism. The reader has become well aware by now of the fact that my pilot beacon in all procedures in non-living nature was what I had learned from the behavior of the basic emotions and the OR energy in the living organism. This principle still holds today and will be valid for a long time to come. I even would venture to suggest to hold on to this principle in studying cosmic energy for all future times. There is no better beacon than the clicking in unison of the inner and outer processes of nature. More, without such harmony little can be expected to be achieved in the realm of cosmic functioning. And it is exactly this principle together with the above mentioned rule of helping nature take care of itself which distinguishes functionalism from mechanistic-mystical approaches to nature. To find the true borderline between artificial measure and self-regulation is the essence of the art of piloting.

It is also the landmark that divides the domains of the dictator from the social guide. The first tells you *what* to do, *where* to go, *how* to go there and, in his extreme forms, he shoots you or burns you and your books if you do not obey his order. The latter lets you choose your own goal and your own way to your goal, which most likely will on the average be the goal of society at large. He tells you: "If you ask me for advice *how* to arrive safely at your destination, *how* to rely on *your own power,* I shall take your hand and guide you. I shall disappear as soon as my job is done."

This basic rule of all true self-regulation, the keeping of proper balance between guidance and free action is in sharp disagreement with both the forceful handling of nature, such as chronic drug supply to the organism on the one hand, and letting the one in need of guidance do as he pleases on the other hand. The desert cannot be dealt with by continuous operations; neither can it be left to itself or be patched up by tedious, costly, frustrating irrigation efforts by way of endless ditching and dam building; if for no other reason than that a dying desert will also sooner or later kill the deep sources of water and destroy the artificial ducts.

There is little use in patching up according to old ways. The discovery of the Life Energy has revealed the manner in which nature works, namely functionally, in a self-regulatory manner, in a steady metabolism between tendency to decay and over-abundance. The problem of natural self-regulation will constitute a special chapter in our account on ORENE, to be published in a different context.

At this point it was important to assume that the desert, once revived in the growing prairie grass, in the decay of the secondary vegetation, in the return of natural manuring by grazing cattle, sheep, horses, etc., in the reinstatement of the natural cycle of sunshine and gentle soaking rain, would tell by itself where the borderline between artificial and self-regulatory management of its affairs could be found. In this process, so much was certain, many still deeply hidden *"obstacles in the way"* of self-regulation would be detected and, if possible, be removed; this would increase the power of self-sustained action and decrease the need for technological interference.

We can imagine this balance between freedom and guidance as being valid for the social forces of human society, too. Here the goal of *work democracy* is to continually reduce the need for governmental or administrative interference and to steadily increase the power of self-management of social groupings by constantly *removing the obstacles in the way to complete self-regulation*. Had our senses not been as much dulled as they actually have been by the forces of evil deceit, organized lying and the Stalinite-Hitlerite Hig rules of suspicion and murder, we would have managed to find our way through and out of the maze of a millennial mess of garbled human affairs. To secure the peace and the freedom and the facilities to get at the **"Obstacle in the Way"** is therefore the basic task of all research and social organization, be it in the combat of poverty, or desert or in the overcoming of gravity.

During the following months it was thrilling to search for the hidden borderline between necessary guidance and self-regulation. I approached the task slowly, carefully, with keen focusing on observation. For example, I would let the Cloudbuster draw moisture from the Pacific Ocean until I felt that the atmosphere was sufficiently saturated. Then I would depress the Cloudbuster for a day or two to see what would happen. In the beginning, from about the middle of December to the beginning of January, the DOR atmosphere would return after only a few hours interruption of drawing operations. Then we would set in with drawing until the atmosphere was moist again. We would then repeat this procedure.

Directions of drawing played some role in finding the borderline. Southwestern and westerly drawings were more effective than northerly (from the desert basin) or from the east. Drawing from the northwest proved later to correlate with drawing from the west. Zenith drawings were always of a crucial nature. But the westerly region seemed to offer the greatest riddles. It yielded the best development of moisture flow and cloud banks. However, after a while it became stale, hard to be mobilized; *something was wrong to the west.* I did not know the nature of the westerly obstacle, nor even its distance or location. But, as the weeks passed by, the impression gained weight that such an obstacle existed to the west. It later turned out to be the **"Barrier"** as the Sierra mountain range west of El Centro. Here is important, *not that* it existed, but *how* it was detected. It will be the subject of a special chapter.

"How" to get at the obstacle to natural self-regulation is always far more important than to try to introduce self-regulatory management into a non-self-regulatory system, be it social, individual or natural. Of such principles of basic research few brought up in today's technological mechanics have any inkling or practical knowledge. One would not mind, and one would even acknowledge the necessity under the circumstances of a certain amount of mechanical regulation of affairs if these blunted minds would not assume the arrogant position of judging what

they do not understand or even try to obstruct in a most obnoxious manner.

It turned out in the course of these explorations that Orur was indispensable in finding the obstacles. It acted like a huge, very sensitive instrument. Without Orur the Barrier in the west would never have been actually found or even broken. Further, more detailed and exact Ea research would not have been possible. Moreover, I doubt whether I would ever have reached the point in my work where I could tell myself: Here, at this precise point the divide is located that separates a sick organism needy of constant help from the healthy organism which can proceed on his own power, even if limping once in a while.

To this method of approach I also owe the final understanding of the "dust storm."

The following map approximately depicts the territory subjected to the test of self-regulation:

Fig. 32. The base of operations at Jacumba, California, west of the *"Barrier"*

Experimental Quest for the Ea Barrier

During the tug of war of January 12th, Ea seemed to have surrounded the Tucson Valley en masse; I conceived of the idea that this may be a routine measure on the part of spacemen to attack a certain region. It was known to me that there existed a sharp borderline between the green mountain ranges on the western slopes of the Sierra and the barren, eastern desert slopes. Was it possible that a *"barrier"* was set up there, which separated the green from the barren land? I decided to find out practically. Two operators went to the west coast 400 miles away, to find out whether such a barrier existed. *It did exist.* They reported from the Sierra divide that the clouds coming in from the *West Coast did not pass but dissipated over the barrier on the dividing range.*

High CPM Indicate "Atmospheric Fever"

I would like to introduce at this point the term: **"Atmospheric Fever."** The term "fever" or "high temperature" has heretofore only been applied to living animal organisms. It indicates a severe reaction expressed in a rise of body temperature to certain kinds of irritation. One speaks in classical medicine of "functional" fever when there seems to be no apparent cause for the rise in temperature. "Functional" in our terms indicates a non-material, non-bacterial disturbance in energy equilibrium. This is unknown to and arbitrarily bypassed by the scientist oriented in atomic chemistry terms only.

Since the basic energy functions in the organism are the same as in the atmosphere, it is legitimate to apply conclusions from the one realm to the other.

OR energy excitation is to the orgonomic view presented by a change from the *foggy* to the *pointed* form of existence. (See Fig. 7d, p. 40.) With this change a motor force of a mechanical nature arises. Under circumstances still greatly unknown such a change may be expressed in terms of higher temperature. Under different circumstances no rise in temperature, but only a motoric force develops. The heat variant is an alternative to the motor variant. We may assume that the same energy change which under one set of circumstances causes mechanical motion, causes under another set of circumstances rise in temperature. Since we are free to drop this assumption whenever we find it inapplicable, we may speculate further: **If the OR energy in its pointed, excited form finds no objects to move mechanically it will cause high temperature of gases or solid substances by inner friction.** Raw as this thought is, it deserves consideration.

The clicks at the Geiger counter are doubtless expressions of single OR energy points charging a vacuum, a grid of an electronic tube or moving the membrane of a mechanical sound amplifier.

We are dealing in such thought operations not with highly technically developed variations of mechanistic electronic brains, but with simple ideas and facts regarding most primitive functions of a physical nature. It seems advisable to avoid the complicated superstructure and to adhere to the simple, even the naive.

In accordance with our assumption the atmospheric OR energy will express its excitation about some dangerous factor in the vicinity by developing the pointed state of existence. It will react accordingly on the GM counter with higher counts. The danger may be the outpouring of atomic radiation in the course of an atomic explosion; it may be the presence in the atmosphere of DOR from the exhausts of spaceships; it may also be DOR rising from our atmospheric energy when our OR envelope dies. We should not become rigid but remain mobile in all such thought operations; in this new realm of knowledge anything may be possible, anything beyond our present knowledge may appear on the scene of inquiry.

When an atomic bomb explodes, a huge amount of nuclear material (NU) suddenly irritates in concentrated form an unprepared, unconcentrated atmospheric Life Energy. On the other hand, *when Oranur is operating a very small amount is irritating a highly concentrated Life Energy.*

In the first case the OR energy falls victim to prostration and decay. In the second case, the OR energy reacts after a brief period of consternation or paralysis with a fierce motor force. It is thus understandable that no high background counts are noticeable beyond a narrow circle around point zero of the atomic explosion. However, in the second case the counts are low at the nuclear material while they soar high to many hundred thousands per minute in the surrounding atmosphere. **Orurization** of the atmosphere presents only a variant of the second case. When we orurize the atmosphere we are causing the atmosphere to react fiercely as in oranur by putting a small amount of NR into one of the Spacegun tubes or by swinging such material to and fro near the Spacegun metal base or pipes. Only the form of the oranur reaction varies in these two cases. The principle is the same.

As we have learned, orurization of the atmosphere cleans out any DOR clouds that may be present. It is immaterial here what happens to the DOR energy; whether it is reconverted into OR energy, whether it changes into water vapor, or is absorbed into the ground or into a well.

DOR disappears in all cases.

On February 15th, 1955 at 09:44 hrs. in the morning during our daily conference, a loud explosion was heard to the north; it was felt by all present on the observation deck. Thereupon the sky was closely watched with binoculars and telescope till 14:00 hrs. We assumed that an atomic bomb had been exploded. At 09:49 the atmospheric OR count, measured with the SU-5 GM Counter was 400-600 CPM. The following chart represents the original protocol written during that operation:

Protocol:
Dictated by Wilhelm Reich, M.D.
Date: February 15, 1955   OROP DESERT Ea

Subject: OROP ORUR At OR I.

Time	Run. min.	CPM	ORUR	Phenomenom
2/15/55				
09:44				Loud explosion heard and felt by WR and 4 others.
09:47				DOR settling down to NW.
09:49	0	400-600		
09:50	1	700-750		
	1½	800-900		
09:51	2	1000-700		
09:52	3	600-700		
09:53	4	800-400-500-600		
09:55	6	500		
09:56	7	500	#1. In (Station.)	
09:57	8	500	"	Vapor trails hold long towards W and NW.
09:58	9	600	"	
09:59	10	400	"	
10:00	11	400	"	
	12	500	"	
	13	600	"	
	14	500	"	Clouds are not dissolving, but DOR settling down heavily.
	15	500	"	ORUR does not counteract in CPM's.
	16	500	"	
	17	400	"	
	18	500	"	
	19	400	"	

Wilhelm Reich
Eva Reich M.D.

p.2

Time	Run. min.	CPM	ORUR	Phenomenom
10:10	21	700-600	Start #1 oscillation	
	22	600-400	"	
	23	450-300	"	
	24	300	"	Clouds arranging in parallel patterns. Pulsatory ORUR more efficient than stationary. Clouds keep stratifying.
	25	300	"	GM off.
10:15	26	400-350	"	Sun has a ring.
	27	350	"	
	28	400-450	"	Sky blue overhead; clouded over and stratified.
	29	500	"	
½ hr.	30	450	"	12 milli-Roentgen.
10:20	1	450-500	"	1 cloud streamer laying along Galactic Plane.
	3	600-700	End #1 (27 min.)	Breeze starts W to E.
	4	650		Clouds and blueness overhead keep well.
	6			DOR settling stronger.
	7	500		Eastern half-ring around sun.
	8			DOR spreading strongly; clouds dissolving; No ORUR.
10:30	11			DOR spreading to NW and W--not towards S or E.
	13	800-650	#2. In (Stationary)	
	14	650	(Oscillating)	Clouds forming again.
	15	550	"	DOR reducing from E towards N.
	16	600-650	"	
	17	600	(Stationary)	Vapor trail holds in N from 30° up to Z.
	18	500	"	DOR on horizon and over Tucson strong. Vapor trail holds all the way. Breeze from E. Depressed Cloudbuster from 30 to 10°. DOR disappears completely from horizon with sweeping.

Wilh. Reich   *William Moise*   *Eva Reich M.D.*   *Robert A. McCullough*

Time	Run. min.	CPM	OR UR	Phenomenom
10:44	23	600	(#2 in)	No DOR.
10:45	24	600	"	Rain expected in S-SW part of U.S., not in middle northern.
	25		"	Drawing low over Tucson and NW.
	28	600-800	"	Cloudbuster up to 45° again, sweeping around.
	29	600	"	
10:51	30		"	
10:52	1		"	
	2	500	End #2. (19 minutes)	Cloudbuster at zenith.
	3	600		DOR coming in again.
	5			DOR coming in strong.
	6	600		
	8	600-700		DOR spreading upward to W and N. Clouds again shrinking. CPM in DOR from N to W horizon. To E and S blueness still prevails. DOR seems to approach from N.
11:00	9			Cloud-free sky in arch from W over N to E.
	13	700		
	15	700-800		
	16	800	Start #3. Stationary	Sweeping low to N-NW.
	17	600	"	Sweeping at 30°. No DOR over Tucson. Blue haze over E and S. North and NW DOR-izing.
11:10	19	500	"	Breeze again from E.
	21	700	"	Sweeping over Tucson on horizon. DOR heavy on northern horizon.
11:21	30		"	
11:22	1	700-800	"	
11:26	5	600-700	End #3 (19 minutes)	

Will Raech

Eva Reich M.D.
Robert A. McCullough
William Moise

13:00 CPM: 250-300; About one-half of the sky to the northwest was free of clouds and the southern half was cloudy.

14:00 CPM: 90 -100; Clouds returning to sky. No DOR around.

Summary:

1 hr., 36 min. draw with òne Cloudbuster.

1 hr., 05 min. ORUR.

Signature of Assistants
Wilhelm Reich M.D.
Robert A. McCullough
William Moise
Eva Reich M.D.

We shall extract here only the highlights. Seven minutes after the explosion was heard, the counts rose to 1000 CPM. If, as we assumed, this had been an atomic explosion somewhere on the military proving grounds in Nevada to the north of us, and the high counts were from the NU outpour, the high atmospheric background count had travelled slower than sound and had reached its peak at our GM counter seven minutes later than the sound wave had reached our ears. This seemed very unlikely.

It appeared much safer to assume that the atmosphere around Little Orgonon was conditioned to react to NU irritation and had reacted to an excitation which travelled through the air mass at the speed of sound. There may have been any other mechanism of transfer at work. We did not know and did not want to waste too much time with theoretical speculations. Far more important it was to act upon this event immediately in an experimental manner. The first wave of high OR energy activity had been released by an unknown event, whatever it may have been. We wished to learn how the atmosphere would behave upon long drawn-out orurization. We orurized the atmosphere one hour and five minutes. We drew at the same time one hour and thirty-six minutes.

The atmospheric excitation was still high with 30˚ CPM at 13:00 hrs.; it was still erratic with 100 CPM at 14:00 hrs., more than four hours later.

In the course of this operation on February 15, 1955 a thought kept coming into my mind which seemed absurd, but was irresistible. If the explosion of nuclear material is due to a rapid, instantaneous change of *secondary* ("after matter") cosmic energy from the resting to the mobile state; if furthermore the atmosphere, sufficiently prepared through repeated orurization would render that atmosphere powerful to resist the assault by NR, a preventive remedy against infestation of the atmosphere with atomic dust would have been found.

I cannot tell at all whether my speculation is sound or not. Neither do I wish to speculate further. But the possibility seemed definitely to be within the realm of the rational. More, with due caution against becoming too speculative, the further thought seemed not too rash that by *creating a higher atmospheric potential than that in the atomic bomb the latter could be rendered useless as a war weapon.*

Since I had no way of proving or disproving my contention, I returned soon, still deeply immersed in these thoughts, to my regular OROP Desert Ea activities. The only thing one could do was to keep the atmosphere clean and strong.

We tried to establish the source of the explosion of February 15th. There were no reports on the radio or in the newspapers. The police reporter of the *Arizona Daily Star* had no record of any explosion that day. The Sheriff's office, too, had heard nothing and had received no report on any explosion that day. Some suggested the explosion may have originated at the Davis Monahan Air Force Base. An inquiry a few days later at the base was answered in the negative. There were no records of an explosion. The safety office knew nothing of an explosion either. Finally a sergeant at the legal officer's office told our operator that all such data was classified and that there was no access to it. Asked whether an important observation connected with the explosion could be relayed to someone, he said there was no one to see. Asked whether this answer was to be taken as a negative one, he said it was to be taken as a *"nothing answer"* since the information was classified.

So we never learned what the explosion had been, what had caused the high counts that day.

During our operation that day, DOR had come down fast and heavy to the west and north; it was absent to the south and east. The atmospheric excitation had reacted to orurization with calming down, i.e., with lowering of the CPM.

The long, drawn-out orurization had had positive effects. The somewhat droughty atmosphere of early February yielded to a rich rain on February 17th at three p.m. It had rained all through from the west coast with 1.44 in. at St. Barbara according to Phoenix station KOPO and L.A. station KFI. The rain was gentle, typical Oranur rain, soaking the ground, a very welcome moisture. The rain moved inland according to our draw from SW and west.

Los Angeles Times 2/18/55

RAINFALL FIGURES

Following is the rainfall, in inches, reported from various Southern California points up to 4 p.m. yesterday. (Normal L.A. rainfall to date 9.37 inches):

Location	Storm	Season	Last Year		Location	Storm	Season	Last Year
Los Angeles	.30	7.49	8.76		Manhattan Bch.	.41	6.39	8.17
Agoura	.51	8.86	10.49		Matilija Dam	1.51	10.66	13.64
Alhambra	.51	8.86	11.06		Mecca	T.	2.47	.61
Altadena	.75	10.25	11.36		Meiners Oaks	1.39	11.18	16.51
Anaheim	.63	7.27	9.10		Modjeska	.90	10.31	12.83
Arcadia	.45	9.77	13.54		Monrovia	.77	10.89	12.59
Avalon	.31	6.76	8.70		Montalvo	.90	7.31	11.82
Azusa	.58	9.03	12.69		Monte Nido	.53	9.62	14.33
Bakersfield	.73	3.56	3.08		Monterey Pk.	.48	9.33	10.27
Banning	.73	9.73	9.33		Moorpark	.79	7.10	11.94
Bardsdale	.66	8.89	11.64		Mt. Baldy	.94	14.42	21.53
Bear Valley	1.24	20.42	24.58		National City	.21	4.53	3.46
Bell	.45	7.88	8.64		Newport Bch.	.45	6.30	6.60
Box Springs Mt.	.60	6.39	7.46		Norco	.37	6.59	6.94
Brea	.68	9.23	9.86		N. Hollywood	.64	7.57	8.06
Buena Park	.55	7.06	7.70		Ojai	1.43	9.31	13.17
Burbank	.56	7.25	8.75		Oakview	1.42	9.57	13.18
Cabazon	.62	7.31	8.79		Oceanside	.26	7.26	9.21
Camarillo	.72	7.90	9.42		Olive	.67	7.52	9.11
Campbell's Sta.	.70	8.29	10.26		Orange	.68	6.27	8.75
Capistrano	.97	6.93	9.99		Oxnard	.75	7.31	11.00
Carlsbad	.66	7.54	7.22		Pacific Palisades	.51		
Cathedral City	T.	2.74	2.15		Palm Spgs.	T.	4.56	3.86
Chatsworth	.65	9.85	11.00		Palos Verdes	.55	6.11	9.79
Chino	.62	8.63	9.90		Pasadena	.64	9.72	11.45
Chula Vista	.18	4.74	4.21		Perris	.52	6.88	6.52
Claremont	.52	9.22	13.21		Piru	.80	8.51	11.20
Colton	.47	7.65	8.65		Placentia	.81	9.01	9.45
Compton	.68	7.90	7.87		Pomona	.71	9.66	12.22
Corona	.65	6.64	7.66		Pt. Hueneme	.61	6.31	10.26
Corona del Mar	.50	5.83	7.72		Port. Bend	.65	6.43	7.16
Coronado	.11	4.49	3.91		Poway	.38	6.75	7.72
Costa Mesa	.51	6.57	6.33		Ramona	.42	6.89	7.83
Covina	.55	7.78	11.04		Redlands	.60	7.94	8.18
Cypress	.50	7.28	9.52		Redondo Bch.	.57	6.42	7.52
Del Mar	.45	5.10	4.76		Rialto	.59	10.08	11.44
Del Rosa	.76	9.57	11.29		Rincon	1.56	7.69	12.60
Duarte	.59	8.50	11.75		Riverside	.56	6.86	7.11
East Highlands	.50	8.22	9.20		Rolling Hills	.58	7.07	8.86
El Monte	.51	9.54	9.72		Saddleback	1.45	15.88	27.18
El Segundo	.46	6.85	8.44		San Bernardino	.58	9.43	10.59
Elsinore	.49	6.34	7.42		San Clemente	.75	6.59	7.48
Encino	.61	9.82	9.99		San Diego	.18	5.06	4.69
Escondido	.47	6.90	9.00		San Dimas	.51	8.65	12.93
Etiwanda	.61	11.02	14.17		San Gabriel	.42	8.53	10.22
Fillmore	.64	8.37	11.37		San Jacinto	.78	7.81	6.17
Fontana	.64	10.44	11.68		San Luis Obispo	.70	13.21	13.47
Foster Park	1.09	9.94	12.84		San Marcos	.65	8.85	10.18
Fresno	.93	7.59	5.67		San Pedro	.50	6.61	8.43
Fullerton	.57	6.76	9.32		Santa Ana	.67	6.96	8.02
Gardena	.60	7.03	9.66		S'nta Ana Val.	1.48	10.65	14.06
Garden Grove	.63	6.94	7.48		Santa B'rb'ra	1.44	11.54	8.84
Glen Avon	.48	6.88	8.22		Santa Monica	.39	7.10	9.20
Glendale	.60	9.01	8.47		Santa Paula	.77	8.07	10.28
Harkel Road	.71	6.18	6.60		Santa Susana	.64	8.62	8.43
Hemet	.65	7.48	6.07		Santiago Dam	.69	6.76	11.20
Hermosa	.46	4.86	6.83		Saticoy	.84	8.71	11.79
Huntington Bch.	.50	5.26	6.43		Sawpit Dam	.65	11.20	14.01
Indio	T.	3.16	.82		Seal Beach	.54	5.22	6.84
Inglewood	.37	7.76	5.90		Shadybrook	.56	7.64	8.55
Irvine	.75	6.43	8.31		Sierra Madre	.64	9.87	12.29
Johnson	.70	5.78	7.48		Silverado Canyon	.90	9.45	15.60
Julian	.50	12.16	12.21		Simi	.81	7.54	8.24
La Canada	.80	9.91	11.83		Somis	.74	7.01	12.46
Lachuza Cyn.	.82	10.45	14.77		S. Pasadena	.44	8.82	10.84
La Crescenta	.68	10.46	16.09		Stanton	.58	6.10	7.01
Laguna Beach	.69	6.77	8.27		Tarzana	.50	7.26	8.60
La Habra	.60	8.06	9.66		Temple City	.46	9.41	10.27
La Jolla	.33	5.11	4.95		Thousand O'ks	.51	7.56	9.12
Lake Arrowh'd	1.50	22.19	26.07		Topanga Can.	.59	11.97	15.55
Lake Hodges	.52	6.88	9.43		Toro Canyon	.59	12.07	15.35
Lake Sherwood	.57	10.50	11.13		Torrance	.63	5.78	6.95
Lake Wohlford	.69	7.22	11.19		Trabucco	.82	9.10	9.42
Lambert	.85	6.53	9.11		Tujunga	.62	9.42	10.08
La Mesa	.19	5.55	6.89		Tustin	.71	7.00	7.19
La Quinta	T.	2.92	.68		Upper Ojai	1.36	11.17	14.39
Las Flores Bell	.65	7.59	12.03		Valley Center	.62	7.30	11.11
Lemon Heights	.76	6.68	9.03		Van Nuys	.65	8.59	7.66
Limestone Cyn.	.70	7.14	12.14		Venice	.26	7.26	9.21
Lomita	.59	6.05	8.33		Ventura	1.04	7.59	8.75
Long Beach	.51	6.85	9.94		Westminster	.68	5.71	7.01
Los Alamitos	.51	6.80	7.41		Wintersburg	.64	5.82	7.59
La Verne	.50	8.81	12.11		Yorba Linda	.78	8.64	10.20
Lynwood	.59	8.45	8.18		Yucaipa	.66	9.01	11.15
					Zuma Beach	.77	6.59	11.09

a. Southern California

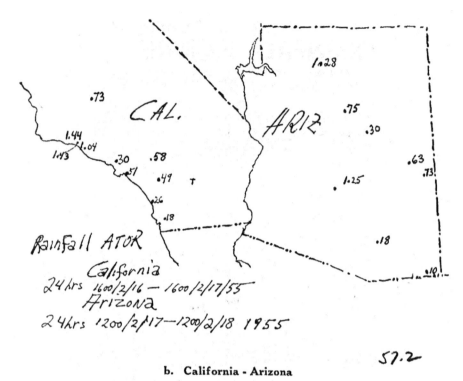

b. California - Arizona

Fig. 33. Rainfall figures, February 16th to 18th, 1955

The clouds were heavy over the Tucson basin on February 17th, but the water refused to fall out of the clouds. I did not understand why. However, with the help of one operator, I succeeded in getting the downpour over Little Orgonon and the whole basin for a duration of three hours. It would be too tedious to report in detail how I got the rain down from the clouds which had refused for hours to pour it out. Roughly speaking, we kept drilling holes into the clouds, alternating with drawings from the still cloud-free sky. We kept irritating the somewhat stale atmosphere until it yielded. It finally did yield and we were happy over our victory. An AAF jet plane flew over Little Orgonon just when the rain began to pour and tipped his wings; they had apparently watched our efforts with interest.

The rain of February 17th had been a fruitful one; the soil in the desert had until then absorbed enough moisture to be well prepared to react to further rain with great productivity. The *Tucson Daily Citizen* wrote on February 23, 1955, page 11 (AP):

> "**Better Range Prospects Seen Because of Winter Storms.**
>
> *Good winter storms promise better range prospects than have been forecast recently.*
>
> The federal crop and livestock reporting service, conducting its monthly survey in Arizona, says cattle are in good shape, although generally affected by the cold.
>
> Conditions by counties: * * *
>
> PINAL—*good spring range forecast due to rains.* Cold weather has held back the spring grass and winter feed is becoming scarce.
>
> PIMA —Supplies of water and range feed adequate, *after a rainfall reported to be the largest in 25 years.*"

As was to be expected from previous experiences, the progress in the combat of desert had also brought about the reaction from the Ea against our efforts. It is naturally so very hard to tell precisely in known terms; the Ea remain mostly invisible and do not behave as any known unfriendly obstacle would do. One of the most reliable criteria to judge extent and nature of Ea activities is to observe the behavior of rainclouds, the strength and frequency of the appearance of DOR clouds, their location and, testing with the Spacegun, their reaction to orurization. It need not be particularly stressed here that these criteria are full of pitfalls. However, as in any other new realm, one becomes accustomed to the new processes; gradually they become familiar and easy to recognize. With this also the safety of judgment increases.

The clouds held well and strong. It had rained all night to the 18th. An A-bomb explosion was predicted by KTUC for 08:30 Mountain Time. A sudden decline in crime by 15% in Los Angeles was reported since January 1, 1955. Had DOR removal and the betterment in the drought situation brought this about? There was no possibility to tell since the relationship to the official administrative centers was, due to the continuous slanderous activities of the various pharmaceutic groups, not as good as it could have been.

We were undecided whether to keep the rain going or not. We phoned Mr. Goyette of the Chamber of Commerce in Tucson. He told us that rain would be harmful from the 22nd to the 28th of February. All Tucson was expecting to enjoy the Rodeo festivals. Thereupon the Cloudbusters were depressed on February 18th, 12:45 hrs. while the drawing for rain was most effective. The manager of the Chamber of Commerce understood full well the nature of our difficulties with the pharmaceutic agents at the University of Arizona; they upheld the interests of the cloud seeders against the interests and wishes of the farmers. The farmers were complaining about the cloud seeders' activities and their financial methods.

To get a message through to the President's office on the gravity of the Ea problem became urgent. For certain reasons it was unwise to press for contacts with the government. Lobbying was going on the bad way to such an extent that the best chance one had in such matters was to be a non-lobbyist. The pressure of Orur and Ea problems grew apace, however.

The Ea were frequent and powerful in the sky again after we had depressed the two spaceguns on February 18th. We could not always tell whether Ea or "the atom" had caused the increasingly frequent high counts in the atmosphere.

The power of Orur was tremendous. It could not be overestimated. Anything seemed possible in its further development. Orur seemed to cause an *extension* of the gaseous atmosphere of the globe. We had no way of proving this impression. However, the jet vapors appeared at the time higher and higher; they seemed to hold better and longer in higher altitudes. One day two jets were seen one above the other. Both poured their vapor into the atmosphere. The lower one held its shape well; the upper one was thin and dissipated quickly. Had the Air Force discovered the extension of the atmosphere? I was not certain at all that this was so. But to judge from the movements of the jets at increasingly higher altitudes admitted of such an assumption. The problem could not be attacked without the facts of Orene.

On February 21, 1955, at nine a.m. the CPM rose again to 300-500 in the atmosphere. It returned to normal at noon. Had an A-bomb explosion, so often postponed lately, been set off? We could not tell. We were all very ill from DOR. The Russians reported a "new remedy against A-blasts." Their statements were never to be trusted since they invented much in their reports according to political expediency. An A-bomb was set off at 06:45 Pacific Time at Nevada. At 08:45 hrs. Tucson Time 300-500 CPM were registered at Little Orgonon. At 10:00 hrs. the counts were 500 CPM. We started orurization of the atmosphere at 10:15; lasting till 10:21. The atmosphere cleared right away. We orurized again for 6 minutes with Orur in lead and a 100,000 +++ CPM reaction followed.

On the 24th of February "radioactive cob-web," a product of Ea in the form of stringy Et, was found in New York state. Severe blackish DOR clouds formed over the Tucson southwestern mountain ranges looking much like tornadoes in formation. They dissolved at 10:35 upon simple drawing. AAF trails persisted well over an hour in the sky that day. At 09:30 hrs. 400 CPM were detected that same morning.

It may well have been due to our daily Orur operation, that the *western barrier became palpable toward the end of February 1955*. It had rained well all night at Oracle Junction 25 miles to the north. The relative humidity again had risen to 70% on the 26th. Phoenix reported moisture streaming in from the south (KUHO weather station). The entire southwest was covered with clouds and had showers. But the desert had only received traces of that rain; 0.09 in. in Phoenix, 0.13 in Douglas, 0.33 in Flagstaff. On the other hand, Hawaii reported 14 inches of rain on the 24th of February. Big floods were reported from Australia. Rain was predicted for practically all of U.S.A. It rained again in southern California, and at Flagstaff and Prescott north of us.

It seemed to rain all around us with the exception of the Tucson basin and the western approaches to Tucson. An Ea was seen on the night of February 26, 1955, in the south low over Tucson at 21:00 hrs. for one hour only. It was not seen later. There were no clouds to the south. It appeared to one operator to be in trouble, flashing in almost pure red three times erratically.

On February 27th, I made a long trip *westward* to inspect the borderline between our experimental and the adjoining territory. Whereas the prairie grass was growing richly to east and north at Oracle and beyond, there was none or only scant grass growing to the west. The line between the region where grass still appeared and the desert with no grass was sharp. Also sharp, ugly, eaten out, black were the rocks on the pass. Sharp DOR, sour taste, mottled palms on the western slope were clear enough indications that a barrier existed to the west, sharply delineated from the clarity of the atmosphere to the east. *I had found the outer reaches of the western barrier.* But it was necessary to find the western beginning of the barrier. This was set into motion during the following days.

On February 28th we measured again 800-1000 CPM with sun, high clouds and 71% humidity. It rained again in California and all over the desert with the exception of Arizona where rain had been predicted but did not come. A flashing Ea had been seen all night to the south over Tucson, 19:45 till about 23:00 hrs. It expired according to operator's report three times. In the west a large orange fiery ball was moving westward at 03:00 hrs. Ea were rather busy in those days.

Probing the Atmosphere on Mountain Ridges

On the night of March 1st I conceived the plan to take the Spacegun toward Oracle and to orurize the atmosphere there. It was necessary to observe the Orur effects from a ridge. For this purpose the mountain ridge at Oracle, 40 miles north of Tucson seemed most suitable since it permitted a view for another 50 miles toward the north, east and west. DOR had been heavy on those faraway mountains most of the time; they were not within the realm of the DOR-free Tucson region. Since, furthermore, I planned to probe the mountain ridge at the tower in the southern Sierra mountain chain, Oracle offered an excellent spot to sharpen one's research tools: *How far would Orur effects reach?*

Though the factual work process developed satisfactorily, the emotional and the social situations were rather distressing. That night all workers in our camp were severely ill from DOR with diarrhea, shivering, vomiting, sleeplessness, thirst, dryness in throat, "agony in the solar plexus," restlessness, vasovegetative reactions in both extremes—paleness and hot flashes. Even our dog, Troll, who was a few months later to become a victim of an evil attack by a hidden Modju, suffered that black day from involuntary bowel movement.

Ea had been there to the south and west for a short while during the night. The severity of DOR-sickness did not seem to depend on the duration of the presence of Ea. It depended on something unknown.

Also, the battle around the discovery of the Life Energy seemed to rage on unabated at that time on the social scene, with our enemies in hiding, working against our expedition and its organizational background like moles in holes. The battle raged at top as well as at local social levels. The readiness to kill by slander, to obtain information by spying, illegal intrusion, etc., or by other equally shabby means was obvious to everyone who partook in those days in our life. The battle was fought in the Weather Bureau, on farms with cattle men and farmers set against the bureaucratic physicists in the local university who protected the chemical interests of cloud-seeding. The drug agents were waiting to break into our files to obtain material both to kill by slander as well as to steal our secrets for the big drug concerns. In the AEC the battle seemed to center around the legitimacy and wisdom of destroying our atmosphere or what was still left of it, by continuous demonstrations of power in exploding atom bombs. The war raged on the ideological scene between those who searched desperately for a peaceful solution of the problem of humanity as did the Eisenhower administration, versus those who continued the age-old murderous system of politics and moral larceny. In this, the western powers, tied down by the rules of democratic procedure, seemed to lose out to the reckless operators without conscience centered in the Stalinite hierarchy. It was to fall later, but it threw its evil power around at that time, supported by equally reckless business pirates in the U. S. A.

I often wondered, and my few assistants with me, how we managed to survive this holocaust at all, with no public support, with a miserably insufficient income from my donation of the medical use of the OR energy of about 30,000 to 40,000 dollars a year, which was threatened at that time by pressure of the drug agency in the government. Only a few low-salaried but devoted men and women held out bravely. Still, somehow the great results we achieved every day in the combat of deserts kept us going.

The atmosphere seemed to be undergoing a radical change. This change was one of the main objectives of our research over the following two months. The Weather Bureau admitted that the rainfall had exceeded the average in the desert region by 1.53 inches. They committed the mistake, however, of deriving their average from long periods, while over several years there had been no rain at all. With respect to the drought of five years, our results were much better than shown by the average.

On March 1st in the early morning an orurization of the atmosphere was conducted for 30 minutes, from 09:05 till 09:35. The background counts that morning were as follows:

09:00	400 - 450 CPM
09:30	600 - 800 CPM
10:00	1000 -1200 CPM
11:30	600 -1500 CPM

The atmosphere fought back the shock with a gentle vehemence. It appeared much more powerful and also successful in keeping development of DOR, i.e., dying OR energy away.

OROP ORUR

Protocol: OROP ORUR, *30 minute ORUR.*

Date: *March 1, 1955*

Place: Little Orgonon, Tucson, Arizona

09:00 No clouds; DOR heavy over Tucson; light breeze from east. At 06:30 in the morning the AEC had exploded a medium sized A-bomb at Yucca Flats, Nevada. *CPM — 450 to 550.*

09:05 ORUR *in;* in pipe, cloudbuster drawing around zenith.

09:07 Breeze from SE. Drawing at a lesser angle. Slowly with steps.

09:10 DOR thinning out fast over Tucson.

09:11 Spiral drawing at about 45°.

09:13 Drawing from around horizon—especially from over Tucson.

09:18 *DOR raising sharply* from over Tucson.

09:20 *500 CPM.* Drawing from just over Catalinas.

09:23 *550 CPM.* Drawing from NW at 45°.

09:26 Drawing from zenith.

09:29 South breeze freshening. *CPM — 600 — 800 — 900 — 1000.*

09:30 *800 CPM.*

09:32 Draw from over Tucson. Feeling of nausea.

09:33 *700 CPM.*

09:35 ORUR *out.* Cloudbuster left at zenith.

Protocol: OROP ORUR, *45 minute ORUR.*

Date: *March 2, 1955*

Place: Little Orgonon, Tucson, Arizona

09:10   High, lacy cloudbank to S, SE and E. Breeze E. Clear overhead. *400 — 700 — 400 CPM.* (WR: Last night it was 150 CPM.) *DOR over Tucson.*

09:15   ORUR *in* (tube). 450 CPM. Drawing around zenith.

09:17   Drawing at 45° to N sector; not S. Vapor trail cohesion increases.

09:20   Clouds increasing, DOR over Tucson moving west, and vapor haze spreading over all of sky (obscuring parts of vapor trails).

09:28   Breeze SE, stronger. Drawing from around zenith. Vapor trails holding.

09:30   *900 CPM.* DOR from Tucson region concentrates over SW mountain region, while remainder of sky clears.

09:33   *500 CPM.* Drawing from NW horizon.

09:34   ORUR out of tube and into lead container on base.

09:40   WR has dizzy spell, ER and WM feel much better than before beginning. Cloudbank to NE and S stretches fingers toward zenith. Sky in haze increasingly obscuring parts of vapor trails. OR flow E to W.

09:44   Drawing from N of Mt. Catalina close to clouds.

09:45   700 CPM. Sweeping E over cloudbank, WM feels this setup stronger than first.

09:50   DOR to SW lifting. Breeze S, strong. Trying to catch breeze.

09:55   Cloudbuster stationary to WNW at 45°.

09:58   *CPM — 500 — 800 — 900*

10:00   Draw from zenith. It doesn't move, try to move it. ORUR *out*. RMC felt it as just past the point where it felt good; a pressure.

10:01   *CPM — 1000 — 1200. WR: there is a point where it shifts over from positive to negative.*

(20 minute ORUR in tube and 25 minute ORUR in container.)

/s/ Robert A. McCullough
Recorder

It appeared not impossible that in due time the atmosphere would be immunized against atomic explosions as one immunizes living systems against infection. Only those ignorant of the functional identity between organismic and atmospheric OR energy would be shocked by such an idea. A highly orurized atmosphere would syphon off any kind of DOR energy from atomic blasts or from Ea. We had the impression that there were difficulties in setting off the A-bomb. That day the riddle of the relationship of DOR to water was solved satisfactorily in the form of an orgonometric equation (see page 258).

On the Way to the West Coast

We were ready to go west to find, explore and if possible to break the barrier. One Orur was to be flown into Jacumba, west of the Sierra divide tower later.

The trip westward some 400 miles was interesting and most rewarding. I saw dry, sandy desert again about 50 miles from Tucson. The contrast to the green prairies which I had just left behind was striking and refreshing since it brought into focus what had been really accomplished in a few months of systematic work.

On the way to Gila Bend, March 3, 1955, 10:30 hrs., 70 miles west of Tucson: there was little DOR. Green grass was sprouting in patches of new vegetation. The soil was moist and fertile looking. For another 30 miles green was still sprouting on both sides of the road; but the vegetation became more sparse. The mountains were bare deserts without any vegetation and desolate looking. *Black rocks,* showing Melanor in heavy attack upon granite and sandstone, were seen about 120 miles west on route 84; 180 miles west of Tucson there appeared spots of a Sahara type desert. Patches of white Et rocks on a black, Melanor infested hill were interspersed with stretches of sand. The static landscape turned into a functional succession of events as it were, revealing its history: Green vegetation, → drought due to DOR attack, → dying of vegetation, → more DOR turning into Melanor on rocks, → eating out rocks, → Melanor turning white, → rocks changing into clay and finally sand.

It seemed incredible that systematic removal of DOR over a few days was sufficient to change the tide of death, to make green life appear again in such desolation. It was still more incredible that this had been unknown to man for ages. Man himself had obstructed the redemption for which he had prayed all over the globe so long, so hard. But then again, I remembered the parents who had worshipped orgonomic knowledge and then had brought their infants to me with high, hardened chests, and pale or blackish cheeks, and how a slight relieving of the high-pressured chest or the arched back had been sufficient to ease the situation, visible in the appearance of a smile and pink color into the face. How patient life is * * *

Near Yuma we passed a stretch of **Black Rocks**. The knowledge of the existence and the qualities of Melanor helped greatly to understand what we saw. First, the evil-looking blackness itself. Now it impressed the onlooker as an attacker, a sapper of strength. It was as if the blackness was eating its way into the rock, causing it to form holes, first small ones, then larger ones. In other places the attacked rock had reached a state where it crumbled into pebbles or rough sand. A whitish hue of Orite had settled over such stretches of initial sand dune development. White Orite was nothing else than the formerly black Melanor. This sounds strange at first, but is easily comprehensible: Melanor, "thirsty" and "hungry" for moisture and oxygen *absorbs* energy only. When saturated it begins to give off or to radiate energy, turning white, entirely in agreement with the absorbing qualities of black and the radiating giving out qualities of white bodies according to Kirchhoff.

255 miles west of Tucson we entered the "American Sahara," vast stretches of full-fledged sand dunes and a rounded, hilly sand landscape. And again the structure of the dunes and hills told the story of their development as the bearing of a human being tells his life story to the knowing eye.

These rounded, smooth sand hills were the ultimate result of the gnawing away of rocks and dry loam by DOR and Melanor. The formerly edgy sharp-cornered turrets had slowly changed into sand dunes. The secondary vegetation had died out, too. And Orite had eaten into the sand to great depth.

Had the huge oil deposits in desert regions anything to do with the penetration of Melanor into the depths? They certainly had. But there is still a long stretch to the full understanding of this riddle.

Could full blown sand deserts be turned into green pastures? This remained to be seen. It seemed unlikely.

At 15:15 hrs. in Grand Wells the gas station man told us that they had had some good rain in the "Sahara" desert in December 1954 and in January 1955, but not before as far as memory could reach. These had been *our* rains; they had come to the "Sahara Desert" with the moisture drawn inland from the west coast. Now I was travelling to inspect the western seashore from where the rain had come to the desert. Finding and destroying the *Barrier* would possibly open the way for more rain. So it happened about one week later.

15 minutes later, after having travelled through the "American Sahara," white Orite was seen richly caking bushes, trees at the roadside: precursors of the complete desolation. How logical it all appeared now, after the discovery of Orene, DOR and Melanor.

On Friday, March 4th, 07:30, 20 miles west of El Centro there was to the northwest a heavy ceiling DOR layer over the mountain range where the barrier was later found and destroyed. Passing through the mountains upward on the road leading to the pass with the outlook tower, we saw now the same DOR ceiling to the east around 08:10 in the morning. It was striking that the higher reaches of the mountains appeared white while the lower ranges were brownish-blackish. Possibly, this was so because Melanor had attacked from above and thus had hit the tops of the mountains first. While satisfying its never-ending thirst for moisture and hunger for oxygen, extracting both from solid granite, and thereby turning white, it attacked the lower regions later and, not saturated yet, it still kept its ugly black color, the physical sign of absorption and greed for moisture.

The tower was reached at 08:45 hrs. We found right away a background count of 400 CPM on the ridge, and 600 CPM three miles to the west at Jacumba. Since high counts always indicated the presence of Ea, there could be no doubts that Ea effects were present over the sharply delineated divide.

On the western side the loam and clay banks showed signs of reorganization toward rock. This was a first sign of a process which ran counter to the desert development. A few miles later we encountered the first green meadows of the western Sierra slopes.

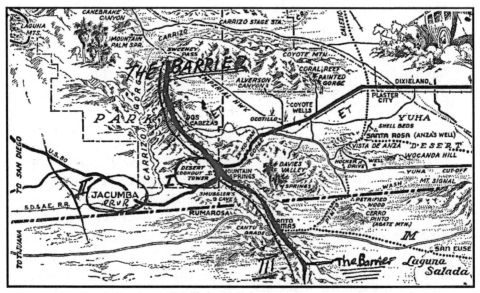

Fig. 34. The "Barrier"
with desert lookout tower where it was broken

The trip down to San Diego was a great pleasure. It was a relief to see green pastures again and a vegetation that was not of the desert type. But the signs of progressing drought were many and unmistakable all through the Pacific Coast. Again, I had the impression of a tug-of-war between the forces of Life and the forces of Death, between OR and DOR, yellow Orene which produced chlorophyll in its merger with the blue atmospheric OR energy and the black Melanor which absorbed the yellow from the vegetation and thus destroyed the green chlorophyll. How far removed from these insights were the mechanistic research methods which were lost in a rigid atomic particle theory, inapplicable to any such natural functions, barren and sterile as was the desert itself, as far as life functions were concerned.

After three delightful days filled with happiness in San Diego, a beautiful, rich city, and after a trip up the coast toward Los Angeles we returned eastward again to the divide and the barrier above it. On the return trip we had the impression still stronger than on the way westward that the green lands were dying. Again the structure of the landscape told the story of a process. The DOR lay heavy in the valleys. The trees were bending more in the valleys than higher up. Meadows and mountain ranges were dying. There were many sandy patches, brownite rocks both decaying and reorganizing.

The Barrier Yields

We arrived at Boulder Park on the divide late that day and made ready to start operations on the barrier the following day, March 6th. I saw that the clouds on approach to or while passing the barrier would dissipate exactly the way I had seen it happen over the observatory at Orgonon in the Oranur years, 1952 and 1953. The whitish, thick clouds would become fuzzy at the advancing fringes, the thin threads would dissolve completely until the whole cloud, no matter how thick would disappear, never to be seen again in its original form. It was absorbed by DOR. This could be seen by its turning in color from white to steel gray or dirty gray and becoming rectangular. Steel gray, rectangular or fuzzy clouds are therefore always signs of the presence of DOR, as are the dirty blackish sky and the high background count. Both indicate effects of present or past presence of Ea. The destruction of the clouds at the western barrier was of this kind and unmistakable.

The truck with the Spacegun came up to the divide at noon on March 6; at 14:00 hrs. operating started. This first operation confirmed fully what we had learned in Maine and in Arizona. The background count was 500 to 800 at 14:00 hrs. There were heavy streaks of blackish DOR across the land in the direction south to north. The clouds were dissipating while passing this barrier. We drew mostly from the zenith for many hours, crisscrossing it; three operators were alternating at the spacegun controls with constant movement of the draw pipes. We did not orurize that day. The result was good. At 17:15 hrs. the zenith was covered with clouds. Clouds began to pass over but were still being dissipated farther to the east. At 17:30 hrs. the background count had come down from the previous 800 to 200 CPM. The following day, on the 7th of March, the counts were 150 at 07:00 hrs., but remained high with 500 to 900 all through the day. We had with our operations apparently drawn some attention from our space visitors.

An atomic bomb was exploded on March 7th, at 05:30 hrs. mountain time. We drew mostly from zenith all that day. The DOR barrier which lay across the mountains, parallel with its north to south direction behaved, to use an example from the realm of the living, like a *"sleeping animal"* that was being tickled; it moved somewhat, but refused to wake up. The air was stale. Upon sufficient drawing, it would stir into a refreshing breeze and the sour taste of DOR would diminish. However, as soon as the operation stopped, the atmosphere returned to its former sluggishness. DOR came down heavy upon the ground. The boulders in the desert to the north which had been gray-white to begin with blackened visibly.

In the early afternoon the barrier seemed to fall apart. Melanor appeared on rocks, ranges, boulders en masse.

The white sands darkened. Around our truck, greening began to appear. We all had the impression that the white sand absorbed the Melanor eagerly. We decided to orurize the following day.

On March 8th, 09:55, the plane arrived from Tucson trailing along orur in the container. The counts inside the plane were 400, 50 feet away 800, and at the container itself 100,000+++.

Operations started at 10:08. The atmosphere gave a count of 900 to 1000 CPM, 12 minutes later 400 CPM. The temperature at the machine had climbed to 110 degrees Fahrenheit. The orur operation was concluded at 11:36 hrs. with a total of 88 minutes. The sky over the former barrier was clear. A fresh breeze was blowing from the west. However, I did not expect that it would last; the barrier was expected to close in again. I decided to leave the Orur material at the Jacumba airport and to send, if necessary, a plane from Tucson for it later on. It remained at Jacumba when the base there was established at a farm.

Three jets with vapor trails were seen between 09:30 and 09:43 hrs. above the barrier, flying from east to west. The vapor trails were short and rather thin.

One operator remained near the Jacumba airport and continued drawing throughout that afternoon after I had left. The response of the atmosphere quickened. The OR energy flow was east to west all day. High thin clouds began to form over the barrier during the night.

At seven o'clock the following morning the clouds still dispersed over the barrier; the atmosphere was blackish and *"flat."* However, the response to the drawing operations was better, with winds coming up from the southwest, always an indication of possible rain. Blueness began to stream into the valleys from the west, the flatness decreased and the atmosphere began to sparkle. The temperature in the sun dropped from 80° to 74° at about 09:00 hrs.

At 11:00 hrs. a blue haze formed in front of the mountains. The OR flow was now very strong and the wind continued from the southwest. Clouds formed rapidly, but still dispersed overhead at the barrier.

At 14:00 hrs. the temperature had dropped to 72°. The atmosphere felt moist. Moisture was streaming in from the southwest. The blue haze and the cloudiness increased further and at 16:00 hrs. the temperature was down to 64°. Most of the time the operator had had contact with the "wind," i.e., the fine breeze which is felt when the operation is in touch with a strong OR stream. The machine was left operating stationary drawing from the southwest. A heavy dew developed during the night. The sky became cloudy. But the clouds were still breaking up. The sky, however, looked whitish and moist.

In the early morning at 08:30 hrs., while "catching the wind," it was found that drawing from the northwest increased the growth of the clouds in zenith and the southwest. At 10:00 the sky was mostly overclouded with thin clouds. Strange cloud formations appeared to the west and moved eastward. These were at first wavy, black lines against the white overcast. They extended downward into streamers, apparently attracted by the still existent DOR ceiling. Then they grew into fat clouds and turned yellow brown. They continued to grow and then became dark blue and heavy like thick rain clouds. At 20:00 the first sprinkle of rain occurred at Jacumba. During the following two hours the rain seemed to have it difficult to come down, but the sky remained covered. The operator continued drawing, as should be done in such cases, from the weak areas in the sky. The cloudbuster was left drawing from low northwest. Light rain began to fall and continued to fall throughout the night where the barrier was breaking up.

The following morning the rain gained strength. It rained at 06:00 hrs. steady and heavily in the typical gentle, soaking Oranur manner. Motorists reported heavy rain in the desert around El Centro across the barrier to the east.

However, at 10:00 hrs. the rain stopped. The clouds began to break up again at the barrier. The operator was unable to stop the break-up with the cloudbuster. At 12:00 hrs. the clouds dispersed at the edge in the southwest. Scattered cloud formations hung over the desert to the east.

The wind continued from the southwest. Also the drawing operations were continued for the sixth consecutive day.

The rain totals on March 11, 1955, were for

Jacumba, Calif.	¼ inch
El Centro	½ inch
San Diego	0.38 inch
La Jolla	0.49 inch
Pasadena	1.00 inch
Los Angeles	0.56 inch
Yuma, Ariz.	traces only

On March 12th there was a heavy dew again during the night. There was, however, only little response to the operation by the atmosphere, possibly because of the moisture having been emptied. The mountain slopes were greening and prairie grass had appeared on the slopes similar to the Tucson region.

The operator returned to base in the afternoon. He drew off and on along the way, too, and he drove with open Cloudbuster pipes.

REPORT ON OROP "ORUR" BARRIER, JACUMBA, CALIFORNIA

Tuesday, March 8th, 1955

Continued drawing throughout afternoon from SW and W and zenith, catching the wind.
Response of atmosphere to cloudbuster gradually quickens during p.m.
Cloudbuster drawing from SW during night, occasional movement.
OR flow was E-W all day.
High thin cloudiness forms during night.
Night warmer than usual.

Wednesday, March 9th, 1955

07:00—No breeze, atmosphere flat, blackish, clouds disperse.
08:00—Began drawing, moving cloudbuster. Response immediate, wind springs up from SW.
Air freshens, blue begins to pour into valley from west.
Flatness decreases and atmosphere begins to sparkle.
OR flow W-E weak.
Temperature in sun drops from 80 at 8 a.m. to 74 by 9 a.m.
11:00—Blue continues to pour in, a blue haze in front of Mts.
OR flow very strong W-E.
Strong wind continues from SW.
Drawing is mostly from SW, Zenith and W.
Clouds begin to form and disperse overhead.
14:00—Temperature in sun down to 72.
Air feels very moist, *as though moisture streaming in from SW.*

15:30—Blue haze and cloudiness increasing.
16:00—Temperature down to 64.
22:00—Cloudbuster now left stationary drawing from SW.
Moving drawing, catching the wind had been conducted most of the time since 8 a.m.

Thursday, March 10th, 1955

Heavy dew during the night. Sky cloudy; overhead and SW, clouds beginning to break up. Sky has a white, moist-looking overcast.
08:30—Began drawing * * * catching the wind * * * found that drawing from the NW increases the cloud growth at Zenith and to SW.
10:00—Sky mostly cloudy, whitish thin clouds * * * strange cloud formations appear to west and move eastward. At first these are wavy, black lines against the white overcast, they send streamers down, begin to grow into fat clouds, turn a yellowish brown, continue to grow and then get very dark blue and heavy * * * like thick rain clouds.
20:00—First sprinkle occurs.
Light short sprinkles occur during next two hours, seems to be having a hard time to come down * * * sky covered. Continued drawing, from weak areas in sky, from low to Mts. to east and from low NW. Low NW seemed to be the most effective.
24:00—Stopped drawing; left cloudbuster drawing from low to NW.
Light rain begins to fall, falls intermittently throughout the night.

Friday, March 11th, 1955

06:00—*Light rain,* increases and rains steady and heavier until 9 a.m. *Rain is soaking and penetrating, steady Oranur type.* Motorists and motorcyclist report heavy rain in desert around El Centro, across the barrier to east.

10:00—Light rain stops, clouds begin to break up * * * unable to prevent breakup with cloudbuster.

12:00—At desert tower * * * clouds coming over from SW expand and disperse at edge * * * scattered cloud formations hang over desert to east.

By drawing from the desert floor from NE, the clouds coming over do not disperse as rapidly or as completely.

Rain totals:

Jacumba, Calif.	¼ inch
El Centro, Calif.	½ inch
San Diego, Calif.	.38 inch
La Jolla, Calif.	.49 inch
Pasadena, Calif.	1.00 inch
Los Angeles, Calif.	.56 inch
Yuma, Arizona	trace

Clouds and Rain came from the southwest.

15:00—Low scattered clouds, very clear, fresh, and clean atmosphere.
Steady wind continues from SW.
Continued drawing from the SW.

Saturday, March 12th, 1955

Heavy dew again during the night. Dew reported to be rare in this area.
OR flow E-W, much blue haze in front of Mts.
Calm, no breeze.
Little response of atmosphere to cloudbuster OROP.
Mountain slopes greening, prairie grass has appeared on the mountain slopes, similar to the Tucson region.

OROP "ORUR", Jacumba, California.

09:00 to 10:00 hrs. (one hour)

Primarily from Zenith and SW * * * attempt to drill holes in atmosphere.

The Weather Bureau had not mentioned the rain over the "American Sahara" and El Centro. The following letter was therefore sent:

Expedition OROP Desert Ea.
Rt. 6, Box 281
Tucson, Arizona

March 15th, 1955

F. W. Reichelderfer
Chief
United States Weather Bureau
Washington, D. C.

Dear Mr. Reichelderfer:

We wish to have on record the following facts and events: On March 7th, 8th, 9th, and 10th, 1955 the Orgone Institute's Expedition OROP Desert Ea. conducted experimental Oranur Weather Control operations at Jacumba, California. These experiments resulted in engineered rainfall throughout Southern California, extending from the coast eastward into the *desert regions* of Southern California. Rain reached Jacumba during the night of March 10th and extended eastward into the desert during the early morning hours of March 11th, 1955. Rainfall continued until midmorning in this area, a period of 6 to 8 hours, with an accumulation at Jacumba of ¼ *of an inch* and ½ *inch in the middle of the desert at* El Centro, California.

The above rainfall records and amounts were obtained from on the spot observation. It was quite puzzling when this unusual desert rainfall was neither mentioned nor reported in any of the California or Arizona weather reports, either newspaper or radio. The rainfall chart of this rain published in the Los Angeles Times of March 12th not only failed to mention the desert rain but omitted entirely, rainfall amounts from the El Centro desert region.

In these times to omit or fail to report such an unusual phenomenon is a very serious matter of national concern.

                Sincerely,

                /s/ William Moise
                Secretary
                Orgone Institute Research Labs.
                Expedition OROP Desert Ea.

ADVISORY COMMITTEE ON WEATHER CONTROL
1128 General Services Administration Bldg.

Washington 25, D. C.

21 March 1955

Mr. William Moise, Secretary
Orgone Institute Research Laboratories, Inc.
Route 6, Box 281
Tucson, Arizona

Dear Mr. Moise:

Please let me thank you for sending us a copy of your letter addressed to Dr. Reichelderfer. We appreciate being informed of your activities.

Sincerely,

/s/ Chas. Gardner, Jr.
CHAS. GARDNER, JR.
Executive Secretary

CHAPTER IX

COSMIC SELF–REGULATION IN OR ENERGY METABOLISM

The breaking of the DOR barrier at the Sierra Nevada east of San Diego had accomplished the full breakthrough of fresh OR energy flow from the west and southwest into the desert basin. During 1955 the self-regulatory metabolic cycle process of OR → DOR energy → Rain → OR energy had taken hold of the desert.

DOR energy in man, animal and plant; DOR energy in the atmosphere; DOR energy in outer cosmic space—all are identical and the products of metabolizing primordial cosmic OR energy. Here is the essence of the manner in which OR energy surrounds and expels DOR energy:

On March 15th, 1955, shortly past 11:00 a.m., the SU-5 Geiger Counter showed outside the building at our base 80,000 to 100,000 CPM. Farther out in the courtyard the GM counter was racing over all ranges toward the extreme 100,000 and far beyond. It reacted erratically. I wondered whether anything had gone wrong with the GM counter; however, in the building I found a near normal count of 50-70. Back in the free atmosphere the counter gave again 100,000 plus, and so it was from now onward over many weeks.

There were many AAF planes overhead. They seemed to crisscross the sky in a nervous, searching manner. Had they, too, discovered the unusual irritation of the atmosphere? Their vapor trails were long and held well over long periods of time. The atmosphere appeared blue-gray, with a haze as seen on beautiful warm summer days.

We drove up to Oracle Junction, 40 miles away. Everywhere the counts went up to 100,000+++ in the free air in an erratic, nervous manner. The counts in the car were everywhere normal, 50-80 CPM, but also erratic, jumpy.

There were many jets in the sky; the trails held well everywhere, except at times in zenith.

Strong south wind gusts stirred up sand, forming dust devils here and there. The clouds were stratified, sharp-edged.

What caused the 100,000+++ counts? The counts were the following during the afternoon that day:

13:00 hrs.	100,000+++	CPM
13:30 hrs.	100,000+++	CPM
14:00 hrs.	100,000	CPM cloudy
15:00 hrs.	40,000	CPM DOR "routed," 12 min. orur to zenith
17:30 hrs.	40,000 to 60,000	CPM
20:00 hrs.	40 to 80	CPM

The following day the counts were:

09:00 hrs.	40,000 CPM
12:00 hrs.	100,000 CPM
16:00 hrs.	100,000 CPM
19:00 hrs.	40 – 60 CPM

An Ea had appeared the evening before in the south and had been drawn from 19:10 till 19:30 hrs. by one operator. Had this Ea done some new damage to our atmosphere: Or were we dealing with an unknown kind of self-sustained chain reaction? In any case, the atmosphere reacted as if highly irritated.

On March 17th at 07:00 hrs. I counted 6000 CPM in the open air. At 09:00, 11:00, 12:00, 13:00 and 16:00 the counts went up rapidly to 100,000+++ again. The milliroentgen counts went up to 20 mr/h. That afternoon the counts were 40,000 to 60,000 at 17:00 hrs. at Oracle and 20,000 at Oracle Road at 18:00 hrs. They came down to

600 to 800 CPM at 18:30 hrs. at Little Orgonon. It had rained at Oracle the night before.

I had wired Dr. Silvert in New York:

> "Please measure hourly daytime and nighttime atmospheric Geiger counts per minutes outside city. Wire result. Counts here one hundred thousand plus for the 6th consecutive day."

His answer was:

> "Hourly measurements 120 miles Montauk overnight and return 1800 March 19 to 1500 March 20 show 30 CPM maximum."

In the evening the riddle dissolved within the framework of the Orgone Theory:

It was quite logical to assume that the surface of the globe developed, in agreement with the identical process in living beings, a shell or an armor between the ground and the ceiling DOR. This armor now was cracking up. The ceiling DOR had been removed earlier. The high counts at the GM counter represented the energy discharge and sudden release of highly pent up energy. OR energy in discharge is always motoric and in the state of pointed existence. The following comparison will be given synoptically to illustrate:

Armored Living Organism	Desert
Life Energy gone stale	Same
DOR sequestration = Armor	DOR-ceiling
	DOR-overground
Dryness, thirst	Dryness, parching
Secondary drives	Secondary vegetation
Defensive, "prickly" character	Prickly growth
Adjustment to life with DOR	Adjustment to desert life
Excitation upon breaking of armor	High CPM counts upon breaking of DOR
Functional fever: OR expels DOR	Same: OR expels DOR (hurricane, tornado, whirlwind, dust devil)

Once more orgonomic theory formation had reached a new level in its search for the CFP of nature: $N=1$.

The high atmospheric counts remained the same over the following days. I stopped all routine operations and measurements on March 24th, 1955. The counts continued to be near normal within the building: low, with a few hundred CPM in the early morning, increasing to 100,000+++ during the day and decreasing again to a hundred in the evening. It was in step with the sun radiation and its variation in the daily cycle.

Would this self-regulatory chain reaction continue to rise or would it stop some day? During 1955 and 1956 I found out more about this self-regulatory, self-cleaning excitation of the atmosphere in various regions of the U.S.A.

The high counts went with strong moisture and a blue-gray haze. The southwest became cooler. Atomic bomb explosions were postponed for the sixth time on March 20th.

It rained over southern California. At 12:50 heavy clouds were coming in from the west: *"Dust Storms."* The dust was with it, of course, since the gusty wind drove the sand ahead of itself. Basically, however, and in full agreement with the orgonomic assumption that mechanical movement is the result of OR energy movement, the sand storm and the clouds turned out to be DOR clouds, hunted by whirling, spiraling, powerful OR energy. And behind the hunted DOR came rain. As soon as the rain started, the CPM came down from 100,000 to 15,000. When the rain stopped again, the counts went up to 100,000+++.

The clouds had to be triggered by drilling holes into the sky to pour out their rain. The water never reached the ground, as if it were absorbed by or changed into something that was not water.

Water was not being absorbed by DOR. *Water* changed *into DOR*. This became quite obvious. It explained the well-known desert phenomenon of rain coming down from clouds, but stopping in the atmosphere, never reaching the ground. **Water changes into DOR**. And, the next thought logically followed: **DOR changes into water, too**. In form of an orgonometric equation:

$$E \looparrowright \begin{cases} OR \looparrowright \begin{cases} H; H; O \looparrowright H_2O; \text{WATER} \\ O; O \looparrowright O_2; \text{Oxygen} \end{cases} \\ DOR \looparrowright \begin{cases} H; H; \looparrowright 2H; \text{Acid Ion} \\ O; O; O \looparrowright O_3; \text{Ozone} \end{cases} \end{cases}$$

Water degenerates into DOR by deterioration of its chemical constituents into acid ions H^+ and Ozone, O_3.

Water arises from DOR by regrouping of the constituents:

$$\underline{DOR} \looparrowright 2H; 3O \looparrowright HOH; (H_2O); O_2 \looparrowright \underline{OR}$$

It may well be found by further research that the oceans of this planet came about by such transformation of *DOR*, that dug deep into the planetary surface rock, into *water*, which then filled the deep eaten-out canyons.

This touches upon the problem of the *creation of our atmosphere*. No more should be said about this at this point, tempting as immediate speculation on the subject may be.

It rained neatly on March 21, 1955. I concluded our operations on March 24th, 1955. We established a base at Jacumba, equipped with two Cloudbusters, a truck and sufficient laboratory equipment. We wound up our affairs during April and started on the way homeward to Orgonon again at the end of April, 1955.

Our job in Arizona was done.

Written during the winter 1955-1956 from Records of the Expedition OROP DESERT Ea, by Wilhelm Reich.

Wilhelm Reich

INDEX

Air Force and Ea, 48ff
Air Force Technical Intelligence Center (ATIC), 78ff
"American Sahara", 240
Arizona Expedition, transportation of cloudbuster, McCullough, 124ff
Arizona Expedition, trip to Arizona, 111ff
Armored living organism and desert, 257
Atmosphere, extension of and ORUR, 231
Atmosphere, probing on mountain ridges, 233
Atmosphere, radical change in, 235
"Atmospheric fever", 217
Atmospheric "ORUR" effect, 23ff
Atmospheric self-regulation, 212ff
Atomic radiation (see NR and NU)
"Atoms for Peace", 171

Baker, Elsworth, M.D., 133, 170
"Barrier", 212ff
"Barrier" yields, 243ff
"Battle of the Universe", 85ff
Black Orene (Oe), 153ff
Black Rocks, 239ff
Breakdown of spacegun operator, 179ff

Cloudbuster, 33
Committee, Advisory, on Weather Control, 253
Common Functioning Principle (CFP) of Nature, 257
Compass deviation and ORUR, 191
"CORE Men", 70
Cosmic OR ocean, 41
Cosmic self-regulation 254ff
Crumbling of rock, 46, 155

"DeDORizing", 115
Desert and armored living organism, 257
"Desert armor", 116

INDEX

Desert development, 148ff
DOR (Deadly Orgone Energy), 148ff
  and disintegration of forests, 152
  and jet vapor trails, 89ff
  and slag from outer space, 155
  and water, 150, 258
  ceiling, 172ff
  clouds, 111
  desert, 121
  desert, functional succession of events, 239
  Emergency, 46
  Emergency observations, 111ff
  erosion of rock, 146
  hunger for nourishment, 151
  hunted by OR, 166
  pocket, 172ff
  sickness, 233
Dust devils, 166

Ea (definition), xxiii
  attack on cloudbuster, McCullough, 139ff
  and CPM, protocol, 221ff
  and stars, differentiation between, 12
  Battle of Tucson, 199ff
  DOR and cloud formation, 165ff
  nodes of, 201
  questionnaire by ATIC on sighting, 55ff
  survey on, 68ff
  swinging, 98
  under ORUR influence, 35, 43, 83ff, 207ff
  "We were being watched", 31
Emotional desert, 151
Equations, orgonometric, 73ff, 99ff, 102ff, 258
Et (Orite), 129, 159
Explosion of Feb. 15, 1955, 221ff

Financial Committee, 133
Fluorescent lights and Orene, 153
"Flying Saucer" problem, 68
Functionalism and mechanism, 98

INDEX

Galactic OR stream, 167
GM, 100,000 CPM and ORUR, 44, 197
GM, 100,000+++CPM, March, 1955, 254
Gravity, negative, 95
Greening of sandy desert, 158ff

Hoff, Grethe, 134

Jet vapor trails and DOR, 89ff

Keyhoe, Donald, Report on UFO's, 7
"Knowledge of the Future", 9
KRW ("Kreiselwellen"), swings ⌒, 15, 71, 95ff

Life energy (Le), 148ff
"Little Orgonon", 143
Lt (dead life energy), 153

McCullough, Robert A., 7, 124ff, 138ff
Mechanism and functionalism, 98
Melanor (Me), 6, 153, 154, 157, 239ff
Metabolism of life energy, 148ff
Moise, William, 78ff, 136

Nature, CFP of, 257
Negative gravity, 95
Nodes of Ea, 201
NR (nuclear radiation), 25
NU (nuclear material), 29, 219

"Oasis", 115
"Obstacle in the Way", 212ff
OR (orgone energy) metabolism and cosmic self-regulation, 254ff
OR motor, 44
OR ocean, cosmic, 41
OR potential, 45, 172
OR stream, galactic, 167
OR units, 37ff
Oranur and NU, 219
Oranur Experiment, First Report, 24, 37ff, 89
Oranur Experiment, Second Report, 25ff

INDEX

Oranur had burst into our lives, 5
Oranur rain, 204ff
Orene (Oe), 152
Orite (Et), 129, 159
OROP Ea, 35
OROP Infant, 115
OROP ORUR, 236ff
OROP "ORUR" Barrier, Report, 248ff
ORUR, 43ff
  and cloudbuster, 33
  and radar, 195
  arrival at Tucson, 183ff
  effect, atmospheric, 23ff
  first operations in desert, 202
  highly excited, 197
  technological use, 28ff
  transfer to Tucson, report, 189ff
  TV report, 195
  vs. nuclear explosions, a possibility, 225
Orurization, 30ff, 219

Pendulum motion, 97
Planetary Professional Citizens Committee, xxiii
"Planetary Valley Forge", 138ff
Proto-vegetation, 158ff

Radar and ORUR, 195
Rain, oranur, 204ff
Rainfall figures, Feb. 16-18, 1955, 227, 228
Rainfall figures, March 11, 1955, 247
"Red Sands", 157
Reich, Ernest Peter, 137
Reich, Eva, M.D., 136, 208ff
Relative humidity rise, 122, 161, 169
"Right to be Wrong", 174ff
Ross, Tom, 185
Ruppelt, E. J., Report on UFO's, xxii, 15, 88

Self-regulation, atmospheric, 212
Self-regulation, cosmic, 254ff
Silvert, Michael, M.D., 114, 133, 189ff

INDEX

"Southern Ea-CM" over Tucson, 208ff
Spacegun, 43
Space War Possible—General MacArthur, xxiii
Spinning waves (see KRW)
Stars and Ea, differentiation between, 12
"Stars" fade out, 3ff
Steig, William, 133
"Swing", ⌢, 95ff
Swing and "Electrocardiogram", 100

T-bodies, 153
Tornadoes, 166
Transfer of ORUR to Tucson, Report, Silvert, 189ff
Transportation of cloudbuster to Tucson, McCullough, 124ff
"Turret" shape in desert, 119

UFO (Unidentified Flying Objects), xxii, 3
UFO, sighting in relation to Venus, 175

"Vacor", 69

White Orene (Oe), 152, 153
White Orite (Et), 129, 159
"White Sands", 158
Work democracy, 214

Yellow Orene, 242

Wilhelm Reich was born in Galicia in the Austro-Hungarian Empire in 1897. After service in the Austrian army in WWI, he attended the University of Vienna Medical School, graduating MD in 1922. He trained in psychoanalysis under Sigmund Freud and from 1924-1929 served as director of the Psychoanalytic Technical Seminar. In addition, from 1922-24 he trained in neuro-psychiatry under Julius Wagner von Jauregg and the Vienna University Neurological and Psychiatric Clinic and worked an additional year in the disturbed wards under Paul Schilder.

Reich introduced many pioneering advances in psychotherapeutic theory and technique, including character analysis, the recognition of character armor and muscular armor, character-analytic vegetotherapy (later called orgone therapy), etc. From 1934-36 Reich conducted a series of laboratory experiments at U. of Oslo Psychological Institute, demonstrating the changes in bioelectric potential of the skin and mucous membranes during differing emotional states. Antithetical changes occurred during sexuality and anxiety. Attempts to carry out similar measurements on microorganisms led serendipitously to origin of life experiments involving bacteria-sized structures Reich called bions.

In one bion culture in 1939, Reich discovered a previously unknown form of radiation. His studies of its properties convinced him it was the specific biological energy hypothesized by B. Moore, P. Kammerer, and others. Reich named this orgone radiation or orgone energy. He found it had potential therapeutic effects on cancer and other diseases. Further research convinced Reich that orgone energy was present in Earth's atmosphere and—still later—throughout the cosmos. The remainder of his scientific career focused on the study of orgone energy physics and astronomy, as well as its behavior in living organisms. Reich emigrated to the United States in Aug. 1939, where he spent the rest of his life.

Reich is also known for pioneering insights in anthropology (*The Invasion of Compulsory Sex-Morality*), in sociology (*The Mass Psychology of Fascism*, and writings on "Work Democracy"), and continued contributions to psychotherapy ("The Masochistic Character," "The Schizophrenic Split," etc.). After a malicious slander article in *New Republic* magazine in 1947, the US Food and Drug Administration began a ten-year legal campaign against Reich, thinly disguised as protecting the public from his experimental device the orgone energy accumulator—but in reality, a concerted effort to destroy Reich's work because some FDA officials perceived it as a sex racket. This campaign succeeded in imprisoning Reich and banning and burning his scientific books and journals in 1956 and 1960. Reich died after eight months of a two-year sentence in federal prison in Nov. 1957. His works were brought back into print beginning in 1960 and are read worldwide up to the present.

Books by **Wilhelm Reich**

American Odyssey

Beyond Psychology

The Bioelectrical Investigation of Sexuality and Anxiety

The Bion Experiments

The Cancer Biopathy

Character Analysis

Children of the Future

Contact with Space

Early Writings

Ether, God, and Devil / Cosmic Superimposition

The Function of the Orgasm

Genitality in the Theory and Therapy of Neurosis

The Invasion of Compulsory Sex-Morality

Listen, Little Man!

The Mass Psychology of Fascism

The Murder of Christ

Passion of Youth

People in Trouble

Record of a Friendship

Reich Speaks of Freud

Selected Writings

The Sexual Revolution

Where's the Truth?

Milton Keynes UK
Ingram Content Group UK Ltd.
UKHW031322010324
438759UK00007B/703